Chat GPT Bible - 10 Books in 1

Everything You Need to Know about AI and Its Applications to Improve Your Life, Boost Productivity, Earn Money, Advance Your Career, and Develop New Skills.

By Codi Byte

Copyright © 2023 by Codi Byte

All rights reserved.

No part of this publication may be reproduced, distributed, or transmitted in any form or by any means, including photocopying, recording, or other electronic or mechanical methods, without the prior written permission of the publisher, except in the case of brief quotations embodied in critical reviews and certain other noncommercial uses permitted by copyright law.

Disclaimer: The information contained in this book is intended solely for educational and entertainment purposes and should not be construed as medical, legal, or professional advice. The author and publisher are not responsible for any actions taken by readers based on the information provided in this book. The reader is solely responsible for their own actions and decisions. The author and publisher do not endorse any specific product, service, or treatment mentioned in this book. The reader should always seek the advice of a qualified professional before making any decisions based on the information provided in this book.

Table of Contents

BOOK 1: ChatGPT for Creativity 1

   Chapter 1: Introduction to ChatGPT and Creativity 2

      The Role of AI in Creativity 2

      Understanding ChatGPT 2

   Chapter 2: ChatGPT and Writing 4

      ChatGPT as a Writing Tool 4

      Enhancing Narrative with ChatGPT 4

   Chapter 3: ChatGPT for Poetry and Prose 6

      Automated Poetry Generation 6

      Prose and Storytelling with ChatGPT 6

   Chapter 4: ChatGPT for Screenwriting and Playwriting 8

      AI in Drama: Screenwriting with ChatGPT 8

      Experimenting with Playwriting and ChatGPT 8

   Chapter 5: ChatGPT and Artistic Inspiration 10

      Stimulating Visual Art Concepts with ChatGPT 10

      Using ChatGPT for Art Critiques and Analysis 10

   Chapter 6: ChatGPT for Music and Sound Design 12

      AI-Generated Song Lyrics 12

      Sound Design Ideas with ChatGPT 13

   Chapter 7: ChatGPT for Innovation and Idea Generation 14

      Using ChatGPT for Brainstorming 14

      Innovation in Business and Technology with ChatGPT 15

   Chapter 8: Ethics of AI in Creativity 17

      Ownership and Originality 18

Bias and Sensitivity in AI-Generated Content 19

Chapter 9: The Future of AI in Creativity 20

Predicted Developments in AI for Creativity 21

Preparing for an AI-Aided Creative Future 22

Chapter 10: Conclusion and Further Resources 24

Reflecting on the Role of AI in Creativity 25

Resources for Further Exploration with ChatGPT 26

BOOK 2: "ChatGPT for Entrepreneurs: Automation and Optimization": Explore how ChatGPT can be used to improve efficiency in the entrepreneurial context. 28

Chapter 1: Introduction to ChatGPT for Entrepreneurs 29

Understanding ChatGPT 30

Importance of AI in Entrepreneurship 31

Chapter 2: Market Research with ChatGPT 33

Identifying Trends with ChatGPT 34

Analyzing Market Data 35

Chapter 3: Idea Generation and Validation 36

Brainstorming Business Ideas with ChatGPT 37

Validating Ideas Using ChatGPT 38

Chapter 4: Business Planning and Strategy 39

Developing Business Plans with ChatGPT 40

Strategic Planning and Forecasting 41

Chapter 5: Using ChatGPT for Marketing and Advertising 43

Content Creation for Marketing 44

Customer Engagement and Interaction 45

Chapter 6: Customer Service Automation 47

Role of ChatGPT in Customer Support 48

    Setting Up Automated Customer Service 49

  Chapter 7: ChatGPT for Product Development 50

    Brainstorming Product Features with ChatGPT 51

    User Experience Design and ChatGPT 51

  Chapter 8: Improving Operational Efficiency 53

    Process Automation with ChatGPT 54

    Optimizing Workflows 55

  Chapter 9: ChatGPT for HR and Team Management 56

    Automated Employee Onboarding 57

    Team Communication and Coordination 57

  Chapter 10: Looking Ahead: ChatGPT and the Future of Entrepreneurship 59

    Current Trends and Future Predictions 60

    Ethical Considerations and Responsible Use of AI in Business 61

BOOK 3: "ChatGPT for Researchers: Assistance in Research and Data Analysis": Shows how ChatGPT can be used as a research and analysis tool. 63

  Chapter 1: Introduction to ChatGPT for Researchers 64

    Understanding ChatGPT 65

    The Role of AI in Research 65

  Chapter 2: Literature Review with ChatGPT 67

    Summarizing Articles with ChatGPT 68

    Using ChatGPT for Comprehensive Literature Searches 68

  Chapter 3: Hypothesis Generation and Validation 70

    Leveraging ChatGPT for Innovative Hypotheses 70

    Validating Hypotheses Using AI 71

  Chapter 4: Research Design and Methodology 73

    Using ChatGPT for Research Design Ideas 73

Evaluating Methodologies with ChatGPT 74

Chapter 5: ChatGPT for Data Collection.. 76

Designing Surveys and Questionnaires 76

Conducting Interviews with ChatGPT 77

Chapter 6: Data Analysis with ChatGPT 79

Qualitative Data Analysis 79

Quantitative Data Analysis 80

Chapter 7: ChatGPT for Writing Research Papers.. 82

Structuring Your Paper with ChatGPT.. 82

Writing Abstracts, Introductions, and Conclusions 83

Chapter 8: Improving Collaboration and Communication... 85

Collaboration in Research Teams 85

Communicating Research Findings to a Non-Expert Audience 86

Chapter 9: Ethical Considerations in AI-Assisted Research 88

Responsible Use of AI in Research 88

Addressing Bias and Fairness 89

Chapter 10: Looking Ahead: ChatGPT and the Future of Research 91

Current Trends and Future Predictions 91

Preparing for an AI-Driven Research Landscape 92

BOOK 4: "ChatGPT for Educators: Tutoring, Feedback and Personalized Instruction": Discusses the use of ChatGPT in the field of education. 94

Chapter 1: Introduction to ChatGPT for Educators... 95

Understanding ChatGPT... 95

The Role of AI in Education 96

Chapter 2: ChatGPT as a Tutoring Tool 98

Reinforcing Classroom Learning with ChatGPT 98

Supplemental Learning Outside of the Classroom 99

Chapter 3: Personalized Instruction with ChatGPT 101

    Tailoring Instruction to Individual Learning Styles 101

    Adapting to Students' Pace and Level of Understanding 102

Chapter 4: Feedback and Assessment 104

    Using ChatGPT for Instant Feedback 104

    Creating and Grading Assessments with AI 105

Chapter 5: ChatGPT in Lesson Planning 107

    Curriculum Design with ChatGPT 107

    Innovative Lesson Ideas and Resources 108

Chapter 6: Encouraging Student Engagement 109

    Using ChatGPT to Facilitate Discussions 109

    Enhancing Student Participation and Collaboration 110

Chapter 7: ChatGPT for Professional Development 112

    Lifelong Learning for Educators 112

    Exploring New Teaching Strategies with AI 113

Chapter 8: Parent-Teacher Communication 115

    Using ChatGPT to Keep Parents Informed 115

    Automating Routine Communications 116

Chapter 9: Ethical Considerations in AI-Assisted Education 118

    Responsible Use of AI in the Classroom 118

    Addressing Equity and Access Issues 119

Chapter 10: Looking Ahead: ChatGPT and the Future of Education 121

    Current Trends and Future Predictions 121

    Preparing for an AI-Driven Educational Landscape 122

BOOK 5: "ChatGPT for Writers: Idea Generation, Correction and Editing": Explores how ChatGPT can

be a powerful tool for writers. **124**

Chapter 1: Introduction to ChatGPT for Writers 125

Understanding ChatGPT 125

The Role of AI in Writing 125

Chapter 2: Idea Generation with ChatGPT 127

Brainstorming Story Ideas 127

Character and Setting Development 128

Chapter 3: Plot Development and Story Arcs 129

Crafting Compelling Plotlines with ChatGPT 129

Building Conflict and Resolution 130

Chapter 4: Writing Dialogue with ChatGPT 132

Creating Authentic Dialogue 132

Improving Character Interaction 133

Chapter 5: Revision and Editing with ChatGPT 135

Proofreading for Grammar and Punctuation 135

Refining Style and Tone 136

Chapter 6: Overcoming Writer's Block with ChatGPT 138

ChatGPT as a Creativity Booster 138

Keeping the Writing Process Flowing 139

Chapter 7: Genre-Specific Writing with ChatGPT 140

Writing for Fiction, Nonfiction, and Poetry 140

Adapting ChatGPT for Different Genres 141

Chapter 8: Writing for Different Mediums 143

Using ChatGPT for Screenwriting, Playwriting, and Blogging 143

Tailoring Content to Specific Mediums 145

Chapter 9: Ethical Considerations in AI-Assisted Writing 146

Plagiarism and Originality 146

Responsible Use of AI in Writing 147

Chapter 10: Looking Ahead: ChatGPT and the Future of Writing 148

Current Trends and Future Predictions 148

Embracing AI in the Writing Process 149

BOOK 6: "ChatGPT for Programmers: Code, Debugging and Optimization": Discusses the use of ChatGPT as a programming assistant. 150

Chapter 1: Introduction to ChatGPT for Programmers 151

Understanding ChatGPT 151

The Role of AI in Programming 151

Chapter 2: Code Generation with ChatGPT 153

Writing Code with AI Assistance 153

Generating Boilerplate and Template Code 154

Chapter 3: Debugging with ChatGPT 155

Identifying Bugs with AI 155

Suggesting Fixes and Improvements 155

Chapter 4: Code Review and Quality Assurance 157

Using ChatGPT for Code Review 157

Ensuring Code Quality with AI 158

Chapter 5: Automated Testing with ChatGPT 159

Writing Test Cases with AI Assistance 159

Automating Test Execution and Reporting 159

Chapter 6: Documentation and Comments 161

Generating Code Documentation with ChatGPT 161

Writing Effective Comments and Docstrings 162

Chapter 7: Optimization and Performance Tuning 163

Identifying Performance Bottlenecks with AI 163

Suggesting Optimization Strategies 163

Chapter 8: ChatGPT for Learning New Programming Languages 165

Using ChatGPT as a Learning Resource... 165

Translating Code Between Different Languages... 166

Chapter 9: Ethical Considerations in AI-Assisted Programming 167

Responsible Use of AI in Coding.. 167

Addressing Bias and Fairness in AI Tools 168

Chapter 10: Looking Ahead: ChatGPT and the Future of Programming... 169

Current Trends and Future Predictions 169

Preparing for an AI-Driven Programming Landscape 169

BOOK 7: "ChatGPT for Professionals: Efficiency, Time Management and Organization": Explores how ChatGPT can be used to improve professional productivity... 171

Chapter 1: Introduction to ChatGPT for Professionals 172

Understanding ChatGPT.. 172

The Role of AI in Professional Productivity 172

Chapter 2: Efficiency and Automation with ChatGPT 174

Automating Routine Tasks 174

Streamlining Workflows with AI 174

Chapter 3: Time Management with ChatGPT 175

Prioritizing Tasks with AI Assistance 175

Scheduling and Calendar Management.. 175

Chapter 4: Organizing Information with ChatGPT 177

Using AI for Data Organization and Retrieval 177

Note-Taking and Meeting Summaries with ChatGPT 177

Chapter 5: Communication and Collaboration 178

Enhancing Team Communication with ChatGPT 178

Using AI for Collaborative Projects and Brainstorming 179

Chapter 6: Professional Learning and Development 180

Utilizing ChatGPT for Skill Development 180

Personalized Learning Plans with AI 181

Chapter 7: Work-Life Balance with ChatGPT 183

Using AI to Manage Personal and Professional Life 183

Mindfulness and Stress Management with ChatGPT 183

Chapter 8: ChatGPT for Remote Work 184

Enhancing Productivity in Remote Settings 184

Adapting to Virtual Collaboration and Communication 185

Chapter 9: Ethical Considerations in AI-Assisted Work 186

Responsible Use of AI in the Workplace 186

Addressing Equity and Access Issues 187

Chapter 10: Looking Ahead: ChatGPT and the Future of Work 189

Current Trends and Future Predictions 189

Preparing for an AI-Driven Work Landscape 190

BOOK 8: "ChatGPT for Social Media Managers: Content Generation, Engagement and Analysis": Shows how ChatGPT can be used in the context of social media. 191

Chapter 1: Introduction to ChatGPT for Social Media Managers 192

Understanding ChatGPT 192

The Role of AI in Social Media Management 193

Chapter 2: Content Generation with ChatGPT 194

Creating Engaging Posts with AI 194

Writing Captions, Tweets, and Other Social Content 194

Chapter 3: Audience Engagement with ChatGPT 196

Automating Response to Comments and Messages 196

Encouraging Interaction with AI 196

Chapter 4: Social Media Analytics with ChatGPT 198

Analyzing Engagement Metrics with AI 198

Understanding Audience Sentiment with ChatGPT 198

Chapter 5: Social Media Strategy and Planning 199

Developing a Social Media Plan with ChatGPT 199

Scheduling and Timing Posts with AI 200

Chapter 6: Crisis Management and Damage Control 201

Using ChatGPT to Monitor Social Media Chatter 201

Responding to Negative Feedback with AI 202

Chapter 7: Brand Voice and Consistency 203

Maintaining a Consistent Brand Voice with ChatGPT 203

Ensuring Content Consistency Across Platforms 203

Chapter 8: ChatGPT for Influencer Marketing 205

Identifying Potential Collaborations with AI 205

Managing Influencer Relationships with ChatGPT 205

Chapter 9: Ethical Considerations in AI-Assisted Social Media Management 207

Responsible Use of AI on Social Media 207

Privacy, Transparency, and Trust in AI Tools 208

Chapter 10: Looking Ahead: ChatGPT and the Future of Social Media Management 209

Current Trends and Future Predictions 210

Embracing AI in the Social Media Landscape 210

BOOK 9: "ChatGPT for Journalists: Research, Writing and Fact-checking": Examines how ChatGPT can be used in the field of journalism. 212

Chapter 1: Introduction to ChatGPT for Journalists 213

Understanding ChatGPT 213

The Role of AI in Journalism 213

Chapter 2: Research and Investigation with ChatGPT 214

Streamlining the Research Process with AI 214

Digging Deeper with Data Analysis 215

Chapter 3: Writing and Editing News Stories with ChatGPT 216

Drafting Articles with AI Assistance 216

Editing and Proofreading with ChatGPT 217

Chapter 4: Fact-checking with ChatGPT 218

Using AI to Validate Information 218

Spotting Fake News and Misinformation 218

Chapter 5: Interviewing and Reporting with ChatGPT 220

Preparing for Interviews with AI 220

Live Reporting and Transcribing with ChatGPT 221

Chapter 6: Data Journalism and Visualization with ChatGPT 222

Analyzing Large Data Sets with AI 222

Creating Data Visualizations with ChatGPT 223

Chapter 7: Social Media and Citizen Journalism 224

Monitoring Social Media Trends with ChatGPT 224

Engaging with the Public and Citizen Journalists 225

Chapter 8: Crisis Reporting and Risk Management 226

Using ChatGPT for Rapid Response Reporting 226

Assessing Risks and Safety in Conflict Zones 227

Chapter 9: Ethical Considerations in AI-Assisted Journalism 228

Responsible Use of AI in Journalism 228

Addressing Bias, Privacy, and Ethics in AI Tools 229

Chapter 10: Looking Ahead: ChatGPT and the Future of Journalism 231

    Current Trends and Future Predictions 231

    Preparing for an AI-Driven Journalism Landscape 232

BOOK 10: "ChatGPT for Linguists: Translation, Interpretation and Language Learning": Discusses the use of ChatGPT for language learning and translation. 235

    Chapter 1: Introduction to ChatGPT for Linguists 236

        Understanding ChatGPT 237

        The Role of AI in Linguistics 237

    Chapter 2: Translation with ChatGPT 238

        Translating Text with AI Assistance 238

        Using ChatGPT for Localization 239

    Chapter 3: Interpretation with ChatGPT 240

        Enhancing Interpretation with AI 240

        Real-time Interpretation with ChatGPT 240

    Chapter 4: Language Learning with ChatGPT 242

        Learning Vocabulary and Grammar with AI 242

        Developing Language Proficiency with ChatGPT 243

    Chapter 5: ChatGPT for Pronunciation and Accent Improvement 244

        Using AI to Improve Pronunciation and Accent 244

        Developing Listening Comprehension Skills with ChatGPT 245

    Chapter 6: ChatGPT for Language Teaching and Assessment 246

        Personalized Learning Plans with AI 246

        Language Assessment and Testing with ChatGPT 247

    Chapter 7: Multilingual Chatbots and Customer Service 248

        Developing Multilingual Chatbots with ChatGPT 248

        Improving Customer Service with AI 249

Chapter 8: ChatGPT for Linguistic Research 250

    Using AI for Linguistic Research and Analysis 250

    Linguistic Data Mining with ChatGPT.. 251

Chapter 9: Ethical Considerations in AI-Assisted Linguistics... 252

    Addressing Bias and Ethics in AI Language Tools.. 252

    Ensuring Privacy and Security in Linguistic Data... 253

Chapter 10: Looking Ahead: ChatGPT and the Future of Linguistics 254

    Current Trends and Future Predictions 254

    Embracing AI in the Linguistic Landscape 255

BOOK 1: ChatGPT for Creativity

Chapter 1:
Introduction to ChatGPT and Creativity

In the vast landscape of modern technology, artificial intelligence (AI) stands as a revolutionary force, driving change in numerous fields. One such field is creativity, where AI's potential has only begun to be tapped. This chapter aims to introduce you to the concept of AI in creativity and familiarize you with ChatGPT, a powerful AI tool that can be harnessed to foster creativity.

The Role of AI in Creativity

Creativity has traditionally been viewed as a uniquely human trait, an intricate fusion of experiences, thoughts, emotions, and the subconscious. However, the advent of AI has broadened this perspective. Today, AI can generate music, paint pictures, write poems, and even come up with innovative ideas. But how does this work?

AI, and specifically machine learning models like ChatGPT, operate by analyzing vast amounts of data and identifying patterns within it. When applied to creative tasks, these models can generate new content that follows the patterns seen in their training data. This could be a new sentence, a piece of music, or a visual artwork. The output is often surprisingly creative, blurring the line between human and machine-generated art

It's important to note that AI doesn't replace human creativity. Instead, it augments it. AI can take care of the more routine parts of the creative process, such as generating initial ideas or editing content. This allows humans to focus on the aspects of creativity that machines can't replicate, like emotion, personal experience, and intricate decision-making.

Understanding ChatGPT

Now, let's delve deeper into ChatGPT. ChatGPT is a language prediction model developed by OpenAI. It's trained on a diverse range of internet text, which equips it with a wide vocabulary and a deep understanding of many topics. When given a piece of text, known as a prompt, ChatGPT generates a relevant and coherent response.

ChatGPT's applications in creativity are vast. In writing, it can be used to generate ideas, draft content, and provide editing suggestions. In art, it can be used to generate descriptions or narratives of visual pieces. For musicians, it can generate song lyrics or suggest melodies. It's an incredibly versatile tool that can enhance creativity in a variety of fields.

The beauty of ChatGPT lies in its interaction with the user. You can guide its responses by carefully crafting your prompts, essentially collaborating with the AI to create something truly unique. It's like having a creative partner who never tires and is always ready to offer new ideas.

In the following chapters, we will explore how to harness the power of ChatGPT in various creative fields. We will provide practical examples, tips, and strategies to help you integrate this powerful AI tool into your creative process.

Remember, AI is just a tool. It's the human wielding the tool who makes the art. And with ChatGPT, you have an exciting new tool to add to your creative arsenal. Let's embark on this journey of AI-enhanced creativity together.

Chapter 2:
ChatGPT and Writing

Writing is a complex, multifaceted endeavor that demands both creative prowess and technical skill. As the digital age unfolds, writers across the globe are leveraging the power of artificial intelligence, specifically language models like ChatGPT, to navigate this demanding craft. In this chapter, we will explore how ChatGPT can serve as a potent writing tool and how it can help enhance narrative creation.

ChatGPT as a Writing Tool

ChatGPT, built by OpenAI, is an AI model trained on diverse internet text. This rich background enables it to understand and generate human-like text, making it an invaluable companion for writers. But how can it help in writing?

First, ChatGPT can be used for ideation. Writers often encounter the infamous "writer's block" - a state of being stuck for ideas or unable to create new work. In such situations, ChatGPT can be prompted to generate ideas or suggest storylines, aiding writers to overcome their creative slump.

Second, ChatGPT can assist in drafting content. It can be prompted to write in a specific style or continue a text snippet in a particular direction. This functionality can help when a writer is struggling to articulate ideas or when they need to write voluminous content in a limited time.

Third, ChatGPT can provide help with editing. It can suggest more fitting words, correct grammatical errors, and even help rephrase sentences for clarity or style. This feature can be instrumental in improving the overall readability and impact of the written piece.

Enhancing Narrative with ChatGPT

Narrative creation, whether it's for a novel, screenplay, or a short story, is a challenging and intricate task. From devising compelling plots to creating engaging characters, there are numerous facets that writers need to juggle. ChatGPT can prove to be a valuable ally in this endeavor.

When it comes to plot creation, ChatGPT can generate a variety of potential storylines based on simple prompts. For instance, if you're writing a mystery novel and are stuck on the plot, you could prompt ChatGPT with "Generate a plot for a mystery novel set in Victorian England". It could also suggest plot twists or help develop subplots, adding depth and complexity to your narrative.

Character creation is another area where ChatGPT can assist. It can help brainstorm character traits, backstories, and even dialogues, providing writers with a more well-rounded understanding of their characters. This could lead to more authentic and relatable characters that resonate with readers.

Furthermore, ChatGPT can help with world-building in genres like fantasy or science fiction. It can generate descriptions of imaginary landscapes, suggest societal structures, or devise fictional languages, helping writers construct rich and immersive universes.

ChatGPT is not just a tool, it's a creative collaborator that can help writers at every stage of their writing process, from ideation to editing. However, it's important to remember that AI is here to augment human creativity, not replace it. The magic of a well-told story still lies in the human touch, the unique perspective that every writer brings to their work. ChatGPT is a powerful resource, but the true power lies in the hands of the writer wielding it. In the next chapters, we will delve deeper into specific applications of ChatGPT in various creative fields.

Chapter 3:
ChatGPT for Poetry and Prose

In the realm of the written word, poetry and prose hold distinct places. They require different stylistic techniques, unique structures, and vary in their rhythmic flow. This chapter aims to shed light on how ChatGPT can be employed to generate both poetry and prose, providing practical tips and techniques along the way.

Automated Poetry Generation

Poetry, with its lyrical aesthetics and profound emotionality, might seem an unlikely domain for an AI. Yet, ChatGPT, with its nuanced language model, can generate surprisingly evocative verses. Here's how you can use ChatGPT for automated poetry generation:

Prompting style and theme: Specify the style of poetry (sonnet, haiku, free verse) and your chosen theme in your prompt. For instance, "Write a sonnet about the changing seasons" or "Compose a haiku about the quiet of midnight".

Emulating poets: If you admire a particular poet's style, prompt ChatGPT to write in that style. For example, "Write a poem in the style of Emily Dickinson about hope."

Completing poems: If you've started a poem but are struggling with how to continue, feed the existing lines to ChatGPT and let it suggest the next ones.

Editing and refining: ChatGPT can also help refine your verses. If you feel a stanza or line is not quite right, ask ChatGPT for alternatives.

Remember, while ChatGPT can generate creative and thematically coherent poetry, the final editing should be done by the human poet to ensure the poem retains a personal touch and emotional depth.

Prose and Storytelling with ChatGPT

When it comes to prose and storytelling, ChatGPT can be a formidable assistant. It can help generate ideas, build characters, devise plots, and even assist with the actual writing process. Here are some practical ways to use ChatGPT for prose and storytelling:

Idea generation: Give ChatGPT a broad theme or genre, and it can generate a host of story ideas. For instance, "Give me ten ideas for a science fiction short story."

Character creation: ChatGPT can help brainstorm characters. For example, "Describe a protagonist for a fantasy novel who is a reluctant hero."

Plot development: ChatGPT can suggest plot twists, conflicts, and resolutions. You could prompt it with something like, "Suggest a plot twist for a detective story where the detective's partner is the main suspect."

Writing scenes: You can use ChatGPT to write entire scenes. For instance, "Write a tense confrontation scene between a rebellious prince and a stern king."

Editing: Feed ChatGPT a paragraph of your prose, and ask it to suggest edits or rewrites. For instance, "How can I make this description more vivid?"

Dialogue generation: ChatGPT can also write dialogues. Give it a scenario and the characters involved, and it can generate a conversation. For example, "Write a dialogue between two astronauts discovering alien life."

ChatGPT can be a valuable tool for both poets and prose writers, offering assistance from ideation to editing. By learning to navigate and use this tool effectively, writers can enhance their creative process, augment their writing skills, and perhaps even break through the dreaded writer's block. In the subsequent chapters, we'll continue our exploration of ChatGPT's applications in other creative domains.

Chapter 4:
ChatGPT for Screenwriting and Playwriting

From the silver screen to the theatrical stage, compelling narratives are at the heart of visual storytelling. The craft of screenwriting and playwriting involves creating riveting plots, compelling characters, and engaging dialogues. This chapter will explore how ChatGPT can contribute to these areas of writing, offering practical tips and techniques for its use.

AI in Drama: Screenwriting with ChatGPT

Screenwriting is a distinct form of storytelling that relies heavily on visual and auditory elements. Here are ways in which you can use ChatGPT to assist with screenwriting:

Concept Generation: A good screenplay starts with a compelling concept. You can prompt ChatGPT to generate unique and interesting concepts for your screenplay. For example, "Generate five high-concept ideas for a psychological thriller."

Scene Creation: ChatGPT can help write individual scenes, complete with dialogue and stage directions. A prompt could be "Write an intense breakup scene in a rainy cafe between a spy and a detective."

Character Development: From creating compelling backstories to developing character arcs, ChatGPT can assist in crafting complex characters. For instance, "Describe the character arc for a soldier suffering from PTSD in a war drama."

Dialogue Generation: Writing convincing dialogue is crucial in screenwriting. ChatGPT can generate dialogue between characters based on the scenario you provide. For example, "Write a witty exchange between a sarcastic superhero and a sardonic villain."

Experimenting with Playwriting and ChatGPT

Playwriting requires a delicate balance of dialogue, action, and characterization to keep the audience engaged. Here are some practical ways to use ChatGPT in playwriting:

Idea Generation: Much like screenwriting, a play begins with a compelling idea. You can prompt ChatGPT to generate unique and creative ideas for your play. For example, "Give me three ideas for a tragicomic play set in a post-apocalyptic world."

Monologue Creation: Monologues are a vital part of many plays. ChatGPT can generate emotional and impactful monologues for your characters. For instance, "Write a monologue for a character who has just lost their twin."

Characterization: You can ask ChatGPT to flesh out your characters, from their personalities to their motives. For example, "Describe a complex antagonist for a political drama set during the French Revolution."

Scene Description: ChatGPT can help with crafting detailed and immersive scene descriptions, setting the stage for your characters. For instance, "Describe the setting for the climactic scene of a play set in a dystopian city."

While ChatGPT does not replace the creative instincts and imagination of a writer, it can certainly aid in the process of crafting narratives for the screen or stage. By providing fresh ideas, writing snippets of dialogue or monologues, and helping with character development, it offers a new avenue for creative exploration in screenwriting and playwriting. In the following chapters, we will continue to explore other creative applications of ChatGPT.

Chapter 5:
ChatGPT and Artistic Inspiration

The crossroads of art and artificial intelligence can lead to some interesting and unexpected destinations. While AI can't paint a masterpiece or sculpt a statue, it can serve as a source of inspiration, stimulating ideas for visual art. Further, it can provide interesting perspectives on art critique and analysis. This chapter delves into these exciting intersections between ChatGPT and the world of art.

Stimulating Visual Art Concepts with ChatGPT

Artists often draw inspiration from a myriad of sources: nature, emotions, experiences, and now, AI. The unique aspect of using ChatGPT as an inspirational tool lies in its ability to combine elements in novel and often surprising ways, pushing the boundaries of traditional thinking. Here's how artists can harness the power of ChatGPT for their creative process:

Idea Generation: When faced with a blank canvas, coming up with an original idea can be daunting. ChatGPT can generate unique concepts based on a given theme or style. For example, "Provide five concept ideas for a surrealistic painting depicting time."

Exploring Symbolism: Symbolism is a powerful tool in art. ChatGPT can suggest symbols to convey particular emotions or themes. You might ask, "What symbols can I use to represent transformation in a sculpture?"

Creating Art Series: Artists often create a series of artworks around a common theme or style. ChatGPT can suggest ideas for such series. For instance, "Suggest themes for a series of abstract expressionist paintings."

Using ChatGPT for Art Critiques and Analysis

ChatGPT's vast training data includes a wealth of art-related content, equipping it to provide intriguing perspectives on art critiques and analysis. This can be useful for art students, critics, and even artists seeking a different viewpoint on their work. Here's how:

Analyzing Art Styles: You can prompt ChatGPT to provide an analysis of different art styles. This could range from broad movements like Impressionism or Cubism to individual artist's styles.

Artwork Critiques: While it's not a substitute for a professional critic, ChatGPT can provide a starting point for critiquing an artwork. You can describe an artwork to ChatGPT and ask for its interpretation.

For example, "Describe the possible meanings of a painting depicting a lone tree in a barren landscape."

Comparing Artworks: ChatGPT can also help compare different artworks, styles, or artists based on the information it has been trained on.

Art History: You can use ChatGPT to explore the context and history of different art movements, styles, and artists. This can provide a deeper understanding and appreciation of the artwork.

Art and AI might seem like strange bedfellows, but as we've seen, there are numerous ways they can interact productively. Using ChatGPT as a tool for stimulating artistic inspiration or for gaining new insights into art critique and analysis can open up unexplored avenues for creativity and understanding. As we move forward, we'll continue to explore how ChatGPT can be a valuable tool in various fields.

Chapter 6:
ChatGPT for Music and Sound Design

An AI language model trained to help with a variety of topics, including music and sound design. I can help you with music theory, composition, production techniques, sound design, and more. Please feel free to ask any questions you may have or provide information about the specific topic you'd like to discuss.

AI-Generated Song Lyrics

AI-generated song lyrics are created using artificial intelligence models, like ChatGPT, that have been trained on a vast amount of text data, including song lyrics from various genres, styles, and time periods. When generating lyrics, the AI model considers factors such as structure, rhyme schemes, and vocabulary commonly found in songs. Here's a step-by-step process for generating song lyrics using AI:

1. Define the theme or topic: Start by specifying the theme, mood, or subject matter for the lyrics. This will help the AI model understand the context and generate more relevant and engaging lyrics.

2. Choose a genre or style: If you have a specific genre or style in mind, let the AI model know. This can help guide the choice of words, phrases, and rhythm to better suit the desired genre.

3. Provide a seed or prompt: You can provide a seed line or phrase to kickstart the AI's creative process. This could be a song title, a specific line, or a series of words related to the theme.

4. Set the structure: Define the structure of your song, including the number of verses, chorus repetitions, and the presence of a bridge or pre-chorus. This will help the AI model create lyrics that fit the desired song format.

5. Generate the lyrics: The AI model will use the provided information to generate lyrics based on patterns and structures it has learned from its training data. It will attempt to create lines that rhyme, have a consistent rhythm, and are thematically relevant to the input.

6. Review and edit: AI-generated lyrics may not always be perfect, so it's essential to review and edit them to ensure they make sense and fit the intended theme, mood, or style. This step may involve rearranging lines, adding or removing words, and adjusting the overall flow of the song.

Keep in mind that AI-generated song lyrics are meant to be a creative tool to help you in your songwriting process. They can provide inspiration, ideas, and a starting point for your lyrics, but it's still up to you to refine and shape them into a finished product.

Sound Design Ideas with ChatGPT

ChatGPT can be a valuable resource for generating sound design ideas and offering suggestions based on your project's needs. Here's a step-by-step process for utilizing ChatGPT to come up with sound design ideas:

1. Define the project: Start by providing information about your project. This may include the type of project (film, game, advertisement, etc.), genre, mood, or any other relevant details that can help set the context for the sound design.
2. Specify the scene or situation: Describe the specific scene or situation that requires sound design. Mention the events taking place, the emotions involved, and any important visual or narrative elements.
3. Explain your goals: Clarify your objectives for the sound design. Are you trying to create a sense of tension, mystery, or excitement? Or maybe you want to convey a specific emotion or atmosphere? Defining your goals will help the AI model provide more relevant suggestions.
4. Request ideas or techniques: Ask ChatGPT for ideas or techniques that fit the context you've provided. You may inquire about specific sound effects, types of music, or sound design techniques that would be appropriate for your project.
5. Evaluate and iterate: Review the ideas and suggestions generated by ChatGPT. You might find inspiration for a particular sound or technique that fits your project well. If you need further clarification or additional suggestions, feel free to provide more information or ask follow-up questions.
6. Implement and refine: Use the ideas generated by ChatGPT as a starting point for your sound design. As you work on your project, you may need to adapt and refine these ideas to better suit your needs and vision.

Remember that ChatGPT is a tool to assist and inspire you in your sound design process. It can provide valuable ideas and guidance, but it's up to you to apply your expertise and creativity to bring those ideas to life in your project.

Chapter 7:
ChatGPT for Innovation and Idea Generation

ChatGPT can be an effective tool for innovation and idea generation across various domains. By providing a starting point or inspiration, ChatGPT can help you brainstorm new concepts, products, services, or even solutions to problems. Here's a step-by-step process for using ChatGPT for innovation and idea generation:

1. Define the problem or domain: Clearly state the problem you are trying to solve or the domain you want to innovate in. Providing context helps the AI model understand your needs and generate more relevant ideas.
2. Specify objectives and constraints: Outline your objectives and any constraints that need to be considered. This may include budget limitations, technical requirements, or market constraints.
3. Request ideas or solutions: Ask ChatGPT for ideas, suggestions, or solutions related to your problem or domain. You may inquire about specific techniques, technologies, or strategies that could be applied to your situation.
4. Evaluate the output: Review the ideas generated by ChatGPT and evaluate their feasibility, relevance, and potential value. Keep in mind that some ideas might be more abstract or require further development, while others might be more concrete and ready to be implemented.
5. Refine and iterate: If you need more information, clarification, or additional ideas, continue the conversation with ChatGPT by asking follow-up questions or providing more details. You can also refine your initial input to narrow down the focus or explore different aspects of the problem.
6. Develop and implement: Use the ideas generated by ChatGPT as a starting point for your innovation process. You may need to conduct further research, consult with experts, or prototype and test ideas to assess their viability.

Remember, ChatGPT is a tool to assist and inspire you in your innovation and idea generation process. It can provide valuable input and help you think outside the box, but it's ultimately up to you to develop and implement these ideas in a meaningful way.

Using ChatGPT for Brainstorming

ChatGPT can be a valuable tool for brainstorming, as it can generate new ideas, help you think outside the box, and explore different perspectives. Here's a step-by-step process for using ChatGPT for brainstorming:

1. Define the topic or problem: Clearly state the topic or problem you want to brainstorm ideas for. Providing context helps the AI model understand your needs and generate more relevant suggestions.
2. Specify objectives and constraints: Outline your objectives and any constraints that should be considered during the brainstorming process. This can include time, budget, resources, or other limitations.
3. Ask open-ended questions: Start the brainstorming process by asking ChatGPT open-ended questions related to your topic or problem. This will encourage the generation of diverse ideas and perspectives.
4. Request multiple ideas: Ask ChatGPT for multiple ideas or solutions to encourage a wider range of possibilities. This can help you identify unique approaches and uncover unconventional solutions.
5. Encourage divergent thinking: Feel free to explore tangential or unrelated ideas that ChatGPT may generate. Sometimes, seemingly unrelated ideas can lead to innovative solutions or spark new connections.
6. Evaluate and refine ideas: Assess the ideas generated by ChatGPT for relevance, feasibility, and potential value. Remember that some ideas may be more abstract and require further development, while others may be more concrete and actionable.
7. Iterate and expand: If you need more ideas or want to explore different aspects of the problem, continue the conversation with ChatGPT by asking follow-up questions or providing additional information. You can also refine your initial input to narrow down the focus or change the direction of the brainstorming session.
8. Organize and prioritize: Once you have a collection of ideas from the brainstorming session, organize and prioritize them based on their relevance, potential impact, and feasibility. This will help you focus on the most promising ideas to develop further or implement.

Remember that ChatGPT is a tool to assist and inspire you in your brainstorming process. It can provide a variety of ideas and help you explore different perspectives, but it's up to you to evaluate, refine, and implement these ideas to achieve your goals.

Innovation in Business and Technology with ChatGPT

ChatGPT can be a valuable resource for generating innovative ideas in business and technology. It can help you brainstorm new concepts, strategies, or solutions to various challenges. Here's a step-by-step process for using ChatGPT for innovation in business and technology:

1. Define the challenge or domain: Clearly state the challenge you are trying to address or the domain you want to innovate in. Providing context helps the AI model understand your needs and generate more relevant ideas.
2. Specify objectives and constraints: Outline your objectives and any constraints that need to be considered. This may include budget limitations, technical requirements, market constraints, or compliance regulations.
3. Request ideas or strategies: Ask ChatGPT for ideas, suggestions, or strategies related to your challenge or domain. You may inquire about specific technologies, business models, or approaches that could be applied to your situation.
4. Explore various perspectives: Encourage ChatGPT to provide diverse ideas and perspectives by asking open-ended questions or requesting alternative solutions. This can help you identify unconventional approaches or uncover hidden opportunities.
5. Evaluate the output: Review the ideas generated by ChatGPT and evaluate their feasibility, relevance, and potential value. Some ideas might be more abstract and require further development, while others might be more concrete and actionable.
6. Refine and iterate: If you need more information, clarification, or additional ideas, continue the conversation with ChatGPT by asking follow-up questions or providing more details. You can also refine your initial input to narrow down the focus or explore different aspects of the challenge.
7. Conduct research and validation: Use the ideas generated by ChatGPT as a starting point for further research and validation. Consult with experts, analyze market trends, or prototype and test ideas to assess their viability and potential impact.
8. Develop and implement: Once you have identified promising ideas, develop a plan to implement them in your business or technology project. This may involve refining the idea, allocating resources, and defining clear goals and milestones.

Remember that ChatGPT is a tool to assist and inspire you in your innovation process. It can provide valuable input and help you explore different perspectives, but it's ultimately up to you to research, develop, and implement these ideas to achieve success in your business or technology endeavors.

Chapter 8:
Ethics of AI in Creativity

The use of AI in creativity raises several ethical questions and concerns. As AI becomes more prevalent in creative fields such as music, art, literature, and design, it's important to consider the ethical implications of its use. Some of the key ethical concerns include:

1. Authorship and ownership: When AI generate creative work, questions arise regarding who should be credited as the author and who owns the rights to the work. Should the AI model be considered the author, or should the credit go to the person who used the AI, the developers of the AI model, or the organization that owns it?
2. Originality and authenticity: AI-generated creative works may blur the lines between originality and imitation, as AI models are often trained on large datasets of existing creative works. This raises concerns about whether AI-generated content is truly original or just a recombination of existing ideas and styles.
3. Copyright and intellectual property: As AI-generated works can sometimes resemble existing creations, it's possible for AI to inadvertently infringe on copyright or intellectual property rights. Determining liability in such cases can be complex and may require new legal frameworks.
4. Bias and fairness: AI models may inadvertently perpetuate biases present in their training data, leading to the creation of biased or offensive content. Ensuring that AI-generated works are fair and unbiased is crucial for maintaining ethical standards in creative industries.
5. Impact on human creativity and employment: The increasing use of AI in creative fields could potentially displace human creators, reducing opportunities for artists, writers, and designers. This raises concerns about the long-term impact of AI on human creativity and employment in these industries.
6. Privacy: The use of AI in creativity may involve processing personal data, such as individual preferences, artistic styles, or creative habits. Ensuring that personal data is handled responsibly and in compliance with data protection regulations is an essential ethical consideration.

To address these ethical concerns, stakeholders in the field of AI and creativity should work together to develop guidelines, best practices, and regulatory frameworks. This may involve encouraging transparency in AI development, fostering collaboration between human creators and AI, and promoting education and awareness about the ethical implications of AI in creative domains.

Ownership and Originality

Ownership and originality are two essential concepts in the creative domain, as they pertain to the rights and authenticity of creative works. Let's explore these concepts in detail:

Ownership: Ownership refers to the possession and control of rights over a creative work. In the context of intellectual property (IP) law, ownership generally means holding copyright or related rights to a work, which grants the owner exclusive rights to reproduce, distribute, perform, display, or create derivative works.

1. Attribution: Properly attributing authorship is an essential aspect of ownership. In traditional creative fields, the individual or group who created the work is considered the author and holds the copyright.
2. Transfer of rights: Ownership rights can be transferred, licensed, or sold. This means that the original creator may not always retain full ownership of their work, depending on contractual agreements and IP laws.
3. Moral rights: In some jurisdictions, authors also have moral rights, which include the right to be recognized as the author of the work and the right to protect their work from distortion, mutilation, or other modifications that may harm their reputation.

Originality: Originality refers to the quality of being new, unique, and not derived from other works. Originality is a crucial criterion for copyright protection, as copyright law generally protects only those works that exhibit a minimum level of creativity and are not mere copies of pre-existing works.

1. Creativity threshold: The threshold for originality varies between jurisdictions but usually requires that a work has been independently created and involves some level of creativity or artistic expression.
2. Derivative works: Works based on or derived from other works may still be considered original if they involve significant creative input from the author. This can include adaptations, translations, or transformations of pre-existing works.
3. Inspiration vs. copying: While creative works may be inspired by other works, there is a fine line between inspiration and copying. To be considered original, a work should not be a direct copy or a substantially similar reproduction of another work.

When it comes to AI-generated creative works, ownership and originality become more complex. AI models are trained on vast amounts of data from existing works, which may raise questions about the originality of their output. Additionally, determining ownership of AI-generated works can be challenging, as it may involve crediting the AI model, its developers, the user, or a combination of these entities. These complexities call for a reevaluation of current IP laws and the development of new frameworks that address the unique challenges posed by AI-generated creative works.

Bias and Sensitivity in AI-Generated Content

Bias and sensitivity are important considerations when using AI-generated content, as AI models are only as good as the data they are trained on. If the data is biased or insensitive, the AI-generated content will reflect those biases and insensitivities. Here are some practical examples and techniques for addressing bias and sensitivity in AI-generated content:

1. Data selection: One way to reduce bias and increase sensitivity in AI-generated content is to carefully select the data used to train the model. This can include selecting diverse data sources and including data that represents a wide range of perspectives and experiences.
2. Data augmentation: Another technique is to use data augmentation to increase the diversity of the data used to train the model. This can include techniques such as oversampling, undersampling, or data synthesis.
3. Algorithmic transparency: It is important to ensure that the algorithms used in AI models are transparent and explainable, so that any biases or insensitivities can be identified and addressed.
4. Human review: Another technique is to include human review and oversight of AI-generated content, to ensure that it is unbiased and sensitive. This can include reviewing the data used to train the model, reviewing the output of the model, and providing feedback to improve the model's performance.
5. Ongoing evaluation: Finally, it is important to continually evaluate and refine AI models to ensure that they are producing content that is unbiased and sensitive. This can include monitoring the performance of the model over time, incorporating feedback from users, and adjusting as needed.

Overall, bias and sensitivity are important considerations when using AI-generated content. By carefully selecting and augmenting the data used to train the model, ensuring algorithmic transparency, including human review and oversight, and continually evaluating and refining the model, businesses can reduce bias and increase sensitivity in AI-generated content.

Chapter 9:
The Future of AI in Creativity

Bias and sensitivity are critical concerns when dealing with AI-generated content, as they can have significant ethical and social implications. AI models, like ChatGPT, are trained on vast amounts of text data from various sources, which may contain biases or sensitive information. Here's a detailed look at these issues and their implications:

1. Bias in training data: AI models can inadvertently learn and perpetuate biases present in their training data. These biases can stem from historical, cultural, or social factors and can manifest in various forms, such as gender, racial, or political biases. Consequently, AI-generated content may exhibit biased language or promote harmful stereotypes.
2. Sensitivity to offensive content: AI-generated content can sometimes include offensive, inappropriate, or harmful language due to biases present in the training data or the model's inability to differentiate between acceptable and unacceptable content. This can lead to content that is offensive or harmful to certain groups or individuals.
3. Filter bubbles and echo chambers: AI-generated content can potentially reinforce existing beliefs or opinions by generating content that aligns with users' preferences, creating filter bubbles or echo chambers. This can limit exposure to diverse perspectives and perpetuate misinformation.
4. Cultural sensitivity: AI-generated content can sometimes lack cultural sensitivity or understanding, leading to content that may be perceived as disrespectful or inappropriate in certain cultural contexts.

To address these concerns and mitigate the risks associated with bias and sensitivity in AI-generated content, the following steps can be taken:

1. Diverse and balanced training data: Ensuring that AI models are trained on diverse and balanced data can help reduce the risk of perpetuating biases. This involves including content from various perspectives, cultural backgrounds, and demographics in the training data.
2. Bias detection and mitigation: Developing techniques for detecting and mitigating biases in AI models can help minimize their impact on generated content. This may involve pre-processing the training data to remove biases, employing debiasing techniques during model training, or implementing post-processing methods to correct biased output.

3. Transparency and explainability: Encouraging transparency in AI development and promoting explainable AI techniques can help stakeholders better understand the sources of biases and sensitivity issues in AI-generated content.
4. User feedback and monitoring: Collecting user feedback and monitoring AI-generated content for bias and sensitivity issues can help identify problematic content and inform ongoing improvements to the AI model.
5. Ethical guidelines and best practices: Developing and adhering to ethical guidelines and best practices for AI-generated content can help organizations navigate the complexities of bias and sensitivity and ensure that the content they produce aligns with societal values and expectations.

Predicted Developments in AI for Creativity

The future of AI in creativity holds significant potential, as advancements in AI technology continue to reshape the creative landscape across various domains. Here are some trends and developments we can expect in the coming years:

1. Collaboration between AI and human creators: As AI becomes more sophisticated, we will likely see more collaboration between AI and human creators. This partnership can lead to unique and innovative creative works, with AI assisting in the ideation, development, and execution of projects.
2. Democratization of creativity: AI has the potential to democratize creativity by providing accessible tools that lower the barriers to entry for various creative fields. This can enable more people to engage in creative pursuits and bring diverse perspectives and ideas to the creative process.
3. Personalization and customization: AI can help create highly personalized and customized content tailored to individual tastes, preferences, and needs. This can lead to more engaging and immersive experiences in fields like gaming, advertising, and storytelling.
4. Enhanced creative tools: AI-powered creative tools will continue to evolve, providing artists, designers, writers, and musicians with more advanced capabilities. This may include AI-assisted design, writing, or composition tools that help streamline and optimize the creative process.
5. Exploration of new creative frontiers: AI can help creators push the boundaries of their creative fields by generating ideas or works that defy conventional norms or explore uncharted territory. This may lead to the emergence of entirely new art forms, genres, or styles.
6. Ethical and legal considerations: As AI becomes more integrated into the creative process, there

will likely be increased attention to ethical and legal issues, such as authorship, ownership, originality, and the potential impact of AI-generated content on society. This may necessitate the development of new legal frameworks and industry guidelines to address these concerns.

7. Education and re-skilling: As AI reshapes the creative landscape, there will be a need for education and re-skilling programs to help individuals adapt to the changing demands of creative industries. This may involve learning to work effectively with AI, mastering new creative tools, or developing new skill sets that leverage the unique capabilities of AI.

In summary, the future of AI in creativity promises exciting opportunities for innovation, collaboration, and democratization. However, it also brings with it ethical and legal challenges that need to be carefully considered and addressed. By embracing the potential of AI while remaining mindful of its implications, we can shape a future where AI plays a positive and transformative role in the creative domain.

Preparing for an AI-Aided Creative Future

Preparing for an AI-aided creative future involves embracing the potential of AI in the creative domain while addressing the ethical and legal challenges that come with it. Here are some steps to help you prepare for this future:

1. Develop AI literacy: Understand the fundamentals of AI, machine learning, and natural language processing. Stay updated on the latest developments, breakthroughs, and trends in AI technology to be aware of its potential impact on your field.

2. Embrace collaboration: Learn to work effectively with AI and view it as a partner rather than a threat. Develop an open mindset and explore opportunities to collaborate with AI in your creative projects.

3. Master AI-powered creative tools: Familiarize yourself with the latest AI-powered tools, platforms, and applications in your creative field. Learn how to effectively use these tools to enhance your creative process and output.

4. Nurture human creativity: Remember that AI is a tool, and human creativity remains central to the creative process. Focus on developing your creative skills, critical thinking, and problem-solving abilities, as these qualities will continue to be highly valuable in an AI-aided creative future.

5. Address ethical and legal challenges: Stay informed about the ethical and legal implications of AI-generated content, such as authorship, ownership, originality, and bias. Educate yourself about intellectual property laws, data protection regulations, and industry guidelines related to

AI in your field.

6. Encourage diversity and inclusion: Ensure that your creative projects are inclusive and represent diverse perspectives. Understand the potential biases in AI-generated content and work proactively to minimize them.
7. Participate in AI-related discussions: Engage in conversations and debates about the role of AI in creativity. Share your thoughts, concerns, and ideas with other creatives, researchers, and policymakers to help shape the future of AI in the creative domain.
8. Adapt and evolve: Be prepared to adapt and evolve in response to the changing creative landscape. This may involve learning new skills, exploring new artistic styles, or embracing new business models that leverage the capabilities of AI.
9. Advocate for responsible AI development: Support initiatives that promote ethical, responsible, and transparent AI development. Encourage the use of AI for social good and help raise awareness about the potential risks and benefits of AI in the creative domain.

By taking these steps, you can prepare yourself for an AI-aided creative future that harnesses the potential of AI while addressing the challenges it presents. Embracing this future will enable you to create more innovative, engaging, and impactful creative works while navigating the evolving ethical and legal landscape.

Chapter 10:
Conclusion and Further Resources

In conclusion, the rise of AI in the creative domain presents a wealth of opportunities and challenges. AI has the potential to reshape the creative landscape by democratizing creativity, enabling new forms of collaboration, and pushing the boundaries of human imagination. However, it also brings ethical and legal concerns related to authorship, ownership, originality, bias, and sensitivity.

To make the most of AI's potential while addressing these challenges, it is essential to stay informed, engaged, and adaptable. Here are some resources to help you further explore the intersection of AI and creativity:

1. Research papers and articles: Keep up-to-date with the latest research and advancements in AI, machine learning, and their applications in creative fields. Websites like arXiv.org and Google Scholar can be useful resources for accessing research papers.
2. Online courses and workshops: Enroll in online courses or workshops on AI, machine learning, and their creative applications. Websites like Coursera, edX, and Udacity offer various courses on AI and related topics.
3. AI-focused conferences and events: Attend conferences, seminars, or webinars that focus on AI and creativity. Events like NeurIPS, ICLR, and SIGGRAPH often feature discussions and presentations related to AI in the creative domain.
4. Newsletters and blogs: Subscribe to newsletters and blogs that cover AI and its impact on creativity. Websites like AIWeirdness, OpenAI, and DeepMind often share updates, insights, and developments related to AI in creative fields.
5. AI-powered creative tools and platforms: Experiment with AI-powered tools and platforms to gain hands-on experience with AI in creativity. Tools like DeepArt.io, RunwayML, and OpenAI's ChatGPT can provide practical insights into AI's capabilities and limitations.
6. Creative communities and forums: Engage with other creatives, researchers, and enthusiasts interested in the intersection of AI and creativity. Online forums, social media groups, and meetups can be excellent platforms for exchanging ideas, discussing challenges, and learning from others.
7. Ethical guidelines and best practices: Familiarize yourself with ethical guidelines and best practices for AI in the creative domain. Organizations like OpenAI, the AI Now Institute, and the Partnership on AI often publish guidelines and recommendations related to responsible AI

development and use.

By staying informed, engaged, and adaptable, you can better prepare for the future of AI in creativity, embracing its potential while navigating the ethical and legal challenges that come with it.

Reflecting on the Role of AI in Creativity

Reflecting on the role of AI in creativity is crucial to understanding its potential impact and the opportunities it presents. Here are some key aspects to consider:

1. AI as a creative collaborator: AI can act as a partner in the creative process, assisting human creators in generating ideas, streamlining workflows, and exploring new creative possibilities. By working alongside AI, human creators can push the boundaries of their creative fields and produce innovative works.
2. Democratization of creativity: AI-powered tools can lower barriers to entry and enable more people to engage in creative pursuits. This can lead to a more diverse and inclusive creative landscape, fostering a broader range of perspectives and ideas.
3. Personalization and customization: AI can help create highly personalized and customized content, catering to individual preferences and needs. This can enhance user experiences across various creative industries, such as entertainment, advertising, and design.
4. Evolution of creative tools: AI will continue to improve and expand creative tools, providing artists, designers, writers, and musicians with advanced capabilities that enhance their creative processes and output.
5. Ethical and legal challenges: The increasing role of AI in creativity raises ethical and legal concerns related to authorship, ownership, originality, bias, and sensitivity. Addressing these challenges will be critical to ensuring that AI is used responsibly and ethically in the creative domain.
6. Impact on human creativity and employment: AI's role in creativity can have both positive and negative implications for human creators. On one hand, AI can augment human creativity, enabling artists to explore new ideas and styles. On the other hand, the widespread adoption of AI in creative fields may lead to concerns about job displacement and the long-term impact on human creativity.
7. Need for adaptability and lifelong learning: As AI reshapes the creative landscape, individuals will need to be adaptable and open to learning new skills and techniques. Embracing AI as a part of the creative process will require continuous learning and re-skilling to stay relevant and competitive in the evolving creative industries.

By reflecting on these aspects, we can gain a deeper understanding of the role AI plays in creativity and its potential to transform the creative domain. Recognizing both the opportunities and challenges presented by AI is essential for navigating the future of creativity and ensuring that AI is harnessed responsibly and ethically.

Resources for Further Exploration with ChatGPT

For further exploration and to make the most of ChatGPT, consider the following resources:

1. OpenAI website (https://www.openai.com): Visit the OpenAI website for detailed information on their projects, research, and AI models, including ChatGPT. The website also contains blog posts, research papers, and updates on the latest developments in AI.
2. OpenAI API documentation (https://platform.openai.com/docs): The OpenAI API documentation provides comprehensive information on how to interact with ChatGPT and other OpenAI models. It offers guidelines, examples, and best practices for developers looking to integrate ChatGPT into their applications.
3. GitHub repositories: Explore GitHub repositories that showcase ChatGPT implementations, tutorials, and open-source projects. These repositories can provide valuable insights and practical examples of how ChatGPT can be used in various applications.
4. Online forums and communities: Engage with other developers, researchers, and enthusiasts in online forums and communities focused on AI and natural language processing. Platforms like Reddit, Stack Overflow, and AI-focused Slack groups can be helpful for discussing ideas, sharing experiences, and learning from others working with ChatGPT.
5. AI conferences and events: Attend conferences, webinars, or workshops that cover AI topics, including natural language processing and generative models like ChatGPT. Events such as NeurIPS, ACL, and EMNLP often feature presentations and discussions related to the latest advancements in AI and their applications.
6. Online courses and workshops: Enroll in online courses or workshops that cover AI, natural language processing, and related topics. Websites like Coursera, edX, and Udacity offer various courses that can help you develop a deeper understanding of AI models like ChatGPT.
7. AI newsletters and blogs: Subscribe to newsletters and blogs that focus on AI, machine learning, and natural language processing. Resources like the AI Alignment Newsletter, AI Weekly, and the Gradient can help you stay updated on the latest developments, breakthroughs, and trends in AI.

By exploring these resources and engaging with the AI community, you can deepen your understanding of ChatGPT, discover its potential applications, and learn how to effectively use it in various contexts.

The best PROMPT for WRITER:

Write in GPT Chat this prompt exactly as it is written below. Then try changing the terms you find in the " " to get the work that works best for you. Remember that in case you are a writer this prompt is PERFECT to give you a story that you then have to articulate further chapter by chapter, you will be amazed at how many famous writers are using this prompt, it is a very powerful tool make good use of it! Prompt:

You are a writer and will act as one. You will write a long story invented by you, with a "fantasy" theme titled "The New Word," for an audience of "fans of the genre," which aims to tell and entertain. The story will have the following structure: Title, Introduction, Chapter 1-2-3-4-5-6-7-8-9-10-Conclusion. You will format the story in markdown. You will use an "exciting" writing style, an "engaging" sentiment, and a "narrative" communicative register.

The best PROMPT for the POET:

Write in GPT Chat this prompt exactly as it is written below. Then try changing the terms you find in the " " to get the work that works best for you. Remember that in case you are a poet this prompt is PERFECT for you, it is a very powerful tool make good use of it! Prompt:

You are a poet and will act like one. You will write a "sonnet" invented by you, with a "patriotic" theme entitled "Let's Make America Great Again", for an audience of "fans of the genre," which aims to tell, entertain and reflect. You will format the poetic manner sonnet. You will use an "exciting" writing style, an "engaging" sentiment, and a poetic communicative register.

The best PROMPT for the FILMAKER:

Write in GPT Chat this prompt exactly as it is written below. Then try changing the terms you find in the " " to get the work that works best for you. Remember that in case you are a filmmaker this prompt is PERFECT for you, it is a very powerful tool make good use of it! Prompt:

You are a filmmaker and will act as one. You will write a long screenplay invented by you, with a "crime" theme titled "The Telephone" for an audience of "fans of the genre," which aims to tell and entertain. The story will have the following structure: Title, introduction, a long development part where you explain how the plot could be developed, and conclusion. You will use an "exciting" writing style, a "suspense" sentiment, and a "narrative" communicative register.

BOOK 2: "ChatGPT for Entrepreneurs: Automation and Optimization": Explore how ChatGPT can be used to improve efficiency in the entrepreneurial context.

Chapter 1:
Introduction to ChatGPT for Entrepreneurs

ChatGPT, developed by OpenAI, is a powerful language model based on the GPT architecture, designed to understand and generate human-like text. Entrepreneurs can leverage ChatGPT to enhance various aspects of their businesses, such as idea generation, content creation, customer support, and more. Here's an introduction to ChatGPT for entrepreneurs:

1. Idea generation and brainstorming: ChatGPT can be an invaluable tool for generating innovative ideas and brainstorming solutions to business challenges. By providing prompts or asking questions, entrepreneurs can use ChatGPT to explore new business concepts, product ideas, marketing strategies, or operational improvements.
2. Content creation: ChatGPT can assist in creating various types of content, such as blog posts, social media updates, marketing copy, product descriptions, or email templates. By providing initial input or guidelines, entrepreneurs can generate high-quality content tailored to their business needs with minimal effort.
3. Market research and analysis: Entrepreneurs can use ChatGPT to help analyze market trends, industry reports, or customer feedback. By asking specific questions, entrepreneurs can gain insights into their target audience, competitors, and potential opportunities or challenges in their market.
4. Customer support: ChatGPT can be integrated into chatbots or customer support systems, enabling entrepreneurs to provide efficient and personalized assistance to their customers. This can help improve customer satisfaction, reduce response times, and free up time for support teams to focus on more complex issues.
5. Task automation: ChatGPT can be used to automate various tasks, such as generating meeting summaries, organizing to-do lists, or managing email correspondence. By automating routine tasks, entrepreneurs can save time and focus on more strategic aspects of their business.
6. Personalized experiences: ChatGPT can help entrepreneurs create personalized experiences for their customers, such as tailored product recommendations or targeted marketing campaigns. By understanding user preferences and needs, ChatGPT can generate content that resonates with the target audience, ultimately driving engagement and conversions.

To make the most of ChatGPT, entrepreneurs should familiarize themselves with the OpenAI API, which provides access to the model and enables integration with various applications. By exploring the potential use cases and benefits of ChatGPT, entrepreneurs can harness the power of AI to enhance their business operations, drive innovation, and create a competitive edge.

Understanding ChatGPT

ChatGPT is an advanced language model developed by OpenAI based on the GPT (Generative Pre-trained Transformer) architecture. It is designed to understand and generate human-like text, making it a powerful tool for a wide range of applications. To understand ChatGPT, let's break down its key components and capabilities:

1. GPT architecture: GPT stands for Generative Pre-trained Transformer. It is a type of AI model that leverages the Transformer architecture, which is a neural network design specifically built for handling sequences of data, such as text. GPT models are pre-trained on vast amounts of text data and fine-tuned for specific tasks.
2. Generative model: ChatGPT is a generative model, meaning it can generate new text based on the input it receives. It does this by predicting the next word in a sequence, given the context provided. This capability makes it suitable for tasks like content creation, text summarization, translation, and more.
3. Contextual understanding: ChatGPT is designed to understand the context of the text it processes. It can capture relationships between words, phrases, and sentences, enabling it to generate coherent and contextually relevant responses. This contextual understanding is what makes ChatGPT a powerful conversational AI.
4. Fine-tuning: Although ChatGPT is pre-trained on large amounts of data, it can be fine-tuned for specific tasks or domains. This allows it to perform well on a variety of applications, including customer support, content generation, and natural language processing tasks.
5. Prompt-based interaction: ChatGPT operates based on prompts, which are text inputs provided by the user. These prompts can be questions, statements, or any other form of text that guides the model's response. ChatGPT generates responses by considering the context of the prompt and its understanding of language.
6. Token-based processing: ChatGPT processes text in chunks called tokens. A token can be a single character or a word, depending on the language. The model's ability to handle a certain number of tokens determines its capacity for processing longer or more complex text.

By understanding these key aspects of ChatGPT, you can gain insight into its potential applications and capabilities. ChatGPT has demonstrated remarkable performance in various natural language processing tasks, making it a valuable tool for businesses, researchers, and developers alike.

Importance of AI in Entrepreneurship

AI has become increasingly important in entrepreneurship as it offers numerous benefits and opportunities for businesses to innovate, scale, and compete in the market. Here are some reasons why AI is crucial for entrepreneurs:

1. Enhanced decision-making: AI can analyze vast amounts of data quickly and accurately, providing valuable insights for better decision-making. By leveraging AI-powered analytics, entrepreneurs can make data-driven decisions that improve business performance and mitigate risks.
2. Improved efficiency: AI can automate repetitive tasks and streamline workflows, freeing up time for entrepreneurs and their teams to focus on more strategic activities. This increased efficiency can lead to cost savings and higher productivity.
3. Personalization and customer engagement: AI enables businesses to create personalized experiences for their customers, tailoring marketing campaigns, product recommendations, and customer support interactions to individual needs and preferences. This helps build customer loyalty, increase engagement, and drive sales.
4. Innovation and product development: AI can assist in generating new ideas, identifying market opportunities, and accelerating product development. Entrepreneurs can leverage AI technologies to create innovative products and services that meet customer demands and differentiate their businesses in the market.
5. Competitive advantage: Early adoption of AI can provide a competitive advantage by enabling businesses to stay ahead of industry trends, respond more quickly to market changes, and deliver superior customer experiences.
6. Enhanced customer support: AI-powered chatbots and support systems can provide fast, accurate, and personalized assistance to customers. This can improve customer satisfaction, reduce support costs, and increase customer retention rates.
7. Scalability: AI can help businesses scale more effectively by automating processes, analyzing data, and optimizing operations. This allows entrepreneurs to grow their businesses more efficiently, even with limited resources.
8. Risk mitigation: AI can help entrepreneurs identify and mitigate potential risks by monitoring real-time data, detecting anomalies, and predicting trends. This can be particularly valuable in areas like cybersecurity, fraud detection, and supply chain management.
9. Talent management: AI can support human resource functions by streamlining recruitment processes, identifying skill gaps, and providing personalized training and development

opportunities for employees.

To harness the full potential of AI in entrepreneurship, it is essential for business owners to stay informed about the latest advancements in AI technology, invest in AI-powered tools and solutions, and develop an AI strategy that aligns with their business objectives. By embracing AI, entrepreneurs can unlock new opportunities for growth, innovation, and success.

Chapter 2:
Market Research with ChatGPT

ChatGPT can be a valuable tool for market research, helping entrepreneurs gain insights into their target audience, competitors, and industry trends. Here are some ways to use ChatGPT for market research:

1. Analyzing customer feedback: You can use ChatGPT to process and analyze customer feedback, such as reviews, social media comments, or survey responses. By asking the model to summarize key points or identify common themes, you can gain insights into customer preferences, pain points, and areas for improvement.
2. Competitor analysis: ChatGPT can help you assess your competitors by generating summaries of their product offerings, marketing strategies, and strengths and weaknesses. You can provide the model with information about your competitors and ask it to compare them to your business or identify potential opportunities for differentiation.
3. Trend analysis: ChatGPT can help you stay informed about industry trends and emerging technologies by processing and summarizing relevant articles, reports, or news stories. By asking the model to identify key trends and insights, you can ensure that your business stays up-to-date and adapts to changes in the market.
4. Identifying target audience: ChatGPT can assist in defining your target audience by analyzing demographic data, customer profiles, or user personas. You can ask the model to identify common characteristics, needs, or preferences among your target customers, helping you tailor your products and marketing efforts accordingly.
5. Generating market research questions: ChatGPT can help you develop a comprehensive market research plan by generating relevant questions to ask during surveys, interviews, or focus groups. By providing the model with your research objectives, you can receive a list of questions designed to gather the necessary insights for your business.
6. Summarizing research findings: After conducting your market research, you can use ChatGPT to summarize your findings and present them in a clear, concise manner. This can help you communicate your insights effectively to stakeholders or team members and inform your strategic decision-making.

While ChatGPT can be a valuable tool for market research, it is important to remember that it may not always provide accurate or up-to-date information, as its training data is limited to what was available up until September 2021. It is crucial to cross-reference and verify the information generated by ChatGPT with other sources to ensure the validity and reliability of your market research findings.

Identifying Trends with ChatGPT

ChatGPT can be a helpful tool for identifying trends in various industries, markets, or topics. To use ChatGPT effectively for trend identification, follow these steps:

1. Define your objective: Clearly outline the goal of your trend analysis, such as identifying emerging technologies in a specific industry or understanding consumer preferences in a particular market.
2. Collect data: Gather relevant data sources, such as articles, reports, news stories, or social media posts related to your topic. You can use this information as input for ChatGPT to ensure it has enough context to generate meaningful insights.
3. Formulate prompts: Design prompts that guide ChatGPT towards providing the information you seek. For example, you can ask the model to "identify key trends in the [industry] over the last two years" or "summarize the most significant changes in consumer preferences for [product category]". Make your prompts as specific and clear as possible.
4. Analyze responses: Review the responses generated by ChatGPT and look for recurring themes, patterns, or insights that can help you identify trends. Be prepared to iterate on your prompts or provide additional context if the initial responses are not satisfactory.
5. Cross-reference and verify: As ChatGPT's knowledge is limited to its training data, which only goes up until September 2021, it is essential to cross-reference the trends it identifies with other sources to ensure their accuracy and relevance. Consult recent reports, articles, or expert opinions to confirm the validity of the trends generated by ChatGPT.
6. Summarize and communicate findings: Once you have identified and verified the trends, use ChatGPT to create a summary of your findings. This can help you effectively communicate the insights to stakeholders, team members, or clients.

By following these steps and leveraging ChatGPT's powerful natural language processing capabilities, you can identify trends that can inform your business decisions, marketing strategies, or product development efforts. Keep in mind that ChatGPT should be used as a supplementary tool, and its outputs should always be verified and complemented by other research methods to ensure a comprehensive understanding of the trends in your area of interest.

Analyzing Market Data

Analyzing market data is an essential aspect of making informed business decisions. While ChatGPT is an excellent tool for text-based analysis, it is not designed specifically for processing and analyzing numerical market data. However, you can still use ChatGPT to assist you in understanding the implications of the data or generating insights based on the information you provide.

Here are some ways to leverage ChatGPT for analyzing market data:

1. Summarize findings: After analyzing market data using specialized tools or software, you can provide ChatGPT with a brief overview of your findings and ask it to generate a concise summary or highlight the most important points.
2. Generate insights: You can provide ChatGPT with your market data analysis results and ask it to generate insights or recommendations based on the data. For example, you could ask, "What are some marketing strategies we can adopt based on the current market trends and customer preferences?"
3. Interpret data: You can use ChatGPT to help interpret specific data points or metrics. For instance, you could ask, "What does a high churn rate imply for our business, and how can we improve customer retention?"
4. Identify patterns or anomalies: Although ChatGPT is not designed to process numerical data directly, you can still provide it with a brief description of any patterns or anomalies you have observed in the data and ask for possible explanations or suggestions on how to address them.
5. Forecasting: While ChatGPT is not specifically designed for forecasting, you can provide it with historical market data and ask for potential future trends or scenarios based on the information given. However, it is crucial to verify and cross-reference these predictions with other forecasting methods or expert opinions.

Remember that ChatGPT's primary strength lies in natural language processing rather than numerical data analysis. To analyze market data effectively, it is essential to use specialized tools or software, such as Excel, Tableau, or Power BI, and consult experts or industry reports to ensure the accuracy and reliability of your findings.

Chapter 3:
Idea Generation and Validation

Idea generation and validation are critical steps in the process of developing new products, services, or strategies. ChatGPT can be a valuable tool in both of these phases. Here's how you can use ChatGPT for idea generation and validation:

Idea Generation:

1. Brainstorming prompts: Start by providing ChatGPT with a prompt that outlines your objectives, such as "Generate innovative product ideas for the sustainable fashion industry." Be specific and clear with your prompts to guide the model towards generating relevant ideas.
2. Explore different perspectives: To generate diverse ideas, ask ChatGPT to consider different perspectives or user personas. For example, "What would be a valuable product feature for a busy professional in the sustainable fashion industry?"
3. Overcome creative blocks: If you're struggling with a particular challenge or aspect of your idea, ask ChatGPT for suggestions or alternative approaches. This can help you overcome creative blocks and discover new solutions.
4. Iterate and refine: Don't hesitate to iterate and refine your prompts to get more detailed or targeted ideas from ChatGPT. You can also provide feedback on the generated ideas to guide the model towards better suggestions.

Idea Validation:

1. Assess feasibility: Once you have a list of ideas, you can use ChatGPT to evaluate their feasibility by asking questions like, "What are the potential challenges in implementing this idea?" or "What resources would be required to bring this idea to life?"
2. Analyze market fit: Use ChatGPT to help you assess the market fit of your idea by asking questions such as, "What are the current trends in the sustainable fashion industry, and how does this idea align with those trends?" or "What is the target audience for this product, and what are their preferences and needs?"
3. Evaluate competition: ChatGPT can assist you in analyzing your competition by asking it to "Identify the main competitors for this product idea and their strengths and weaknesses."
4. Estimate potential impact: Ask ChatGPT to help you estimate the potential impact of your idea by considering factors like customer benefits, potential revenue, or environmental impact.

5. Gather feedback: You can use ChatGPT to generate questions for surveys, interviews, or focus groups to gather feedback from potential users, stakeholders, or experts in the industry.

While ChatGPT can be a valuable tool for idea generation and validation, it is essential to remember that its knowledge is limited to its training data, which only goes up until September 2021. Always verify the information generated by ChatGPT with other sources and consult experts or conduct additional research to ensure the accuracy and relevance of your findings.

Brainstorming Business Ideas with ChatGPT

ChatGPT can be a helpful tool for brainstorming business ideas, as it can provide you with creative suggestions, insights, and perspectives. To effectively brainstorm business ideas with ChatGPT, follow these steps:

1. Define your objective: Clearly outline the goal of your brainstorming session, such as identifying business opportunities in a specific industry or developing a unique product or service idea.
2. Formulate prompts: Design prompts that guide ChatGPT towards providing relevant and useful ideas. For example, you can ask the model to "generate unique business ideas for the healthcare industry" or "suggest innovative solutions to reduce plastic waste." Make your prompts as specific and clear as possible.
3. Explore different angles: To generate diverse ideas, ask ChatGPT to consider various perspectives, user personas, or market segments. For example, "What are some business ideas that cater to the needs of remote workers?" or "What are some sustainable business opportunities in the food and beverage industry?"
4. Iterate and refine: Don't hesitate to iterate on your prompts or provide additional context if the initial responses are not satisfactory. You can also give feedback on the generated ideas to guide the model towards better suggestions.
5. Evaluate and shortlist ideas: Review the ideas generated by ChatGPT and identify those with the most potential. Consider factors such as market demand, feasibility, competition, and alignment with your skills, resources, and goals.
6. Further research and validation: Once you have shortlisted some promising business ideas, conduct additional research and consult industry reports, experts, or potential customers to validate and refine your ideas.

Remember that ChatGPT should be used as a supplementary tool for brainstorming and that its outputs should always be complemented by other research methods to ensure a comprehensive understanding of the business opportunities and challenges. By leveraging ChatGPT's powerful natural language processing capabilities, you can generate creative business ideas that can inspire your entrepreneurial journey and help you discover new market opportunities.

Validating Ideas Using ChatGPT

While ChatGPT can be a useful tool for idea validation, it is essential to keep in mind that its knowledge is limited to its training data, which only goes up until September 2021. Nonetheless, you can use ChatGPT to help you evaluate your ideas from various perspectives and generate relevant questions or considerations. Here's how you can use ChatGPT for idea validation:

1. Assess feasibility: Ask ChatGPT questions related to the feasibility of your idea. For example, "What are some potential challenges in implementing a subscription-based meal kit service?" or "What resources would be required to develop a virtual reality fitness app?"
2. Analyze market fit: Use ChatGPT to help you understand the market fit of your idea. You can ask questions like, "What are the current trends in the online education sector, and how does a gamified learning platform fit within those trends?" or "What is the target audience for an eco-friendly cleaning product, and what are their preferences and needs?"
3. Evaluate competition: To analyze your competition, you can ask ChatGPT questions like, "Who are the main competitors for a plant-based meal delivery service, and what are their strengths and weaknesses?" or "How can we differentiate our AI-powered chatbot from existing solutions in the market?"
4. Identify potential risks: Ask ChatGPT about possible risks or challenges associated with your idea, such as regulatory hurdles, market saturation, or technology barriers.
5. Estimate potential impact: You can ask ChatGPT to help you estimate the potential impact of your idea by considering factors like customer benefits, potential revenue, market growth, or social and environmental impact.
6. Gather feedback: Use ChatGPT to generate questions for surveys, interviews, or focus groups to collect feedback from potential users, stakeholders, or experts in the industry.
7. Refine your idea: Based on the insights and feedback generated by ChatGPT, refine your idea by identifying areas for improvement, optimizing features, or addressing potential challenges.

Remember that ChatGPT should not be your only source of validation for your ideas. It is crucial to cross-reference the information and insights generated by ChatGPT with other sources, such as industry reports, expert opinions, and customer feedback, to ensure the accuracy and relevance of your findings. By combining ChatGPT's capabilities with other research methods, you can validate your ideas more effectively and make informed decisions about your business, product, or service.

Chapter 4:
Business Planning and Strategy

ChatGPT can be a helpful tool in assisting with various aspects of business planning and strategy. Here's how you can leverage ChatGPT for this purpose:

1. Market analysis: Use ChatGPT to analyze market trends, customer preferences, and competition. Ask questions like, "What are the current trends in the renewable energy market?" or "What are the main competitors for an online coaching platform?"
2. Identifying target audience: You can ask ChatGPT to help you define your target audience based on your product or service offering. For example, "What are the characteristics of the target audience for a luxury skincare brand?"
3. SWOT analysis: ChatGPT can help you conduct a SWOT (Strengths, Weaknesses, Opportunities, and Threats) analysis of your business. Provide the model with information about your business and ask it to identify the key aspects of your SWOT analysis.
4. Setting objectives: ChatGPT can help you set specific, measurable, achievable, relevant, and time-bound (SMART) objectives for your business. For example, "What are some SMART objectives for a new e-commerce store selling eco-friendly products?"
5. Brainstorming strategies: Use ChatGPT to generate potential strategies for achieving your business objectives. For example, "What are some marketing strategies to increase the online visibility of our mobile app?" or "What are some cost-effective ways to improve customer retention?"
6. Risk assessment: ChatGPT can help you identify potential risks or challenges associated with your business plan or strategy. Ask questions like, "What are some potential risks in expanding our business internationally?" or "What are the possible challenges of adopting a remote work model for our company?"
7. Developing action plans: Ask ChatGPT to help you create action plans to implement your strategies or achieve your objectives. For example, "What are the steps to launch a successful crowdfunding campaign?" or "How can we optimize our supply chain to reduce costs and improve efficiency?"
8. Financial planning: While ChatGPT is not specifically designed for financial analysis, you can still use it to generate questions or considerations for your financial planning, such as "What are the key financial metrics to track for our SaaS business?" or "What are some strategies to improve

cash flow management?"

Keep in mind that ChatGPT's knowledge is limited to its training data, which goes up until September 2021. Always cross-reference the information generated by ChatGPT with other sources and consult experts or conduct additional research to ensure the accuracy and relevance of your findings. By leveraging ChatGPT's capabilities and combining them with other research methods, you can create more informed and effective business plans and strategies.

Developing Business Plans with ChatGPT

ChatGPT can be a helpful tool in developing business plans by providing insights, guidance, and prompts to facilitate the process. Here's how you can use ChatGPT to develop a business plan:

1. Define your business: Provide ChatGPT with a brief description of your business, its objectives, and its target audience.
2. Analyze the market: Use ChatGPT to help you understand the market trends, customer preferences, and competition in your industry. Ask questions like, "What are the current trends in the e-commerce market?" or "What are the main competitors for a healthy food delivery service?"
3. Identify your unique value proposition: Ask ChatGPT to help you identify and articulate your unique value proposition that sets you apart from your competitors. For example, "What is the unique selling point of a sustainable fashion brand?"
4. Create a SWOT analysis: Use ChatGPT to help you create a SWOT (Strengths, Weaknesses, Opportunities, and Threats) analysis of your business. Provide the model with information about your business and ask it to identify the key aspects of your SWOT analysis.
5. Develop a marketing strategy: Ask ChatGPT to help you develop a marketing strategy that aligns with your business objectives and target audience. For example, "What are the best marketing channels to reach the target audience for a luxury travel company?"
6. Estimate financial projections: While ChatGPT is not designed for financial analysis, you can use it to generate questions or considerations for your financial projections, such as "What are the key revenue streams for a software-as-a-service (SaaS) business?" or "What are some common expenses to consider for a small retail business?"
7. Define operational processes: Ask ChatGPT to help you define the operational processes, such as supply chain management, logistics, or customer service, that are critical for your business's success.
8. Set milestones and objectives: Use ChatGPT to help you define specific, measurable, achievable,

relevant, and time-bound (SMART) objectives and milestones for your business plan.
9. Iterate and refine: Review and refine your business plan with ChatGPT's insights and feedback. Iterate on your prompts or ask for additional questions to optimize your plan's effectiveness.

Remember that ChatGPT should not be your only source of information or guidance for developing your business plan. It is crucial to cross-reference the insights and recommendations generated by ChatGPT with other sources, such as industry reports, expert opinions, or customer feedback, to ensure the accuracy and relevance of your findings. By leveraging ChatGPT's capabilities and combining them with other research methods, you can create more comprehensive and effective business plans that align with your objectives and target audience.

Strategic Planning and Forecasting

Strategic planning and forecasting are critical components of any business's success. ChatGPT can be a useful tool to assist with these tasks by providing insights, data analysis, and projections. Here's how you can use ChatGPT for strategic planning and forecasting:

1. Analyze market trends: Use ChatGPT to help you understand the current and future market trends in your industry. Ask questions such as "What are the emerging technologies in the renewable energy sector?" or "What are the potential market opportunities in the e-learning industry?"
2. Forecast demand: ChatGPT can help you forecast the demand for your product or service by considering factors such as market size, customer behavior, and competition. Ask questions such as "What is the expected demand for our product in the next three years?" or "How can we anticipate changes in consumer behavior in response to the pandemic?"
3. Identify potential risks: Ask ChatGPT to help you identify and assess potential risks or challenges that may impact your business in the future. For example, "What are the risks associated with expanding our business internationally?" or "What are the potential regulatory changes that may impact our industry?"
4. Develop scenarios: ChatGPT can assist you in developing different scenarios for your business based on various assumptions and variables. For instance, "What would be the potential impact on our revenue if we increase our advertising spending by 50%?" or "What would be the best-case and worst-case scenarios for our sales in the next year?"
5. Evaluate opportunities: ChatGPT can help you evaluate the potential opportunities for your business by considering various factors such as market demand, competition, and resources. Ask questions such as "What are the potential benefits of entering a new market segment?" or

"What are the opportunities for diversifying our revenue streams?"

6. Forecast financials: ChatGPT can help you generate financial projections based on your assumptions and data inputs. You can ask questions like "What would be the projected revenue and expenses for the next three years?" or "What would be the return on investment (ROI) for a new product launch?"

7. Develop strategies: Based on the insights and data generated by ChatGPT, you can develop strategies and action plans to achieve your objectives and respond to potential risks or opportunities.

Remember that ChatGPT's knowledge is limited to its training data, which goes up until September 2021. Therefore, it is crucial to cross-reference the information generated by ChatGPT with other sources and consult experts or conduct additional research to ensure the accuracy and relevance of your findings. By leveraging ChatGPT's capabilities and combining them with other research methods, you can develop more informed and effective strategic plans and forecasts that align with your business objectives and target audience.

Chapter 5:
Using ChatGPT for Marketing and Advertising

ChatGPT can be a helpful tool in assisting with various aspects of marketing and advertising by providing insights, suggestions, and content ideas. Here's how you can use ChatGPT for marketing and advertising:

1. Identify your target audience: Use ChatGPT to help you define and understand your target audience based on their demographics, behavior, and preferences. For example, "What are the characteristics of the target audience for a premium skincare brand?" or "What are the preferences and needs of the target audience for a new vegan protein bar?"
2. Develop brand messaging: Ask ChatGPT to help you develop clear and compelling brand messaging that resonates with your target audience. For example, "What are some key value propositions for a sustainable fashion brand?" or "What are some effective taglines for a new line of organic baby products?"
3. Generate content ideas: Use ChatGPT to generate content ideas for your marketing and advertising campaigns, such as blog posts, social media posts, or video ads. For example, "What are some engaging Instagram captions for a new line of fitness apparel?" or "What are some creative ways to promote a new eco-friendly cleaning product?"
4. Analyze customer feedback: Ask ChatGPT to help you analyze customer feedback, such as reviews, ratings, or social media comments, to identify areas for improvement or opportunities to enhance your marketing and advertising efforts.
5. Optimize advertising campaigns: ChatGPT can help you optimize your advertising campaigns by generating suggestions for targeting, ad copy, or design. For example, "What are some effective ad targeting options for a new line of vegan pet food?" or "What are some best practices for designing a banner ad for a health and wellness app?"
6. Monitor brand reputation: Use ChatGPT to help you monitor your brand reputation online by identifying potential brand reputation issues, such as negative reviews or social media mentions. For example, "What are some effective ways to respond to negative customer reviews?" or "How can we proactively address potential reputation risks on social media?"
7. Evaluate campaign effectiveness: Ask ChatGPT to help you evaluate the effectiveness of your marketing and advertising campaigns by generating questions or considerations related to metrics such as reach, engagement, conversion rates, or return on investment (ROI).

Remember that ChatGPT should not be your only source of information or guidance for your marketing and advertising efforts. It is crucial to cross-reference the insights and recommendations generated by ChatGPT with other sources, such as industry reports, customer feedback, or marketing experts, to ensure the accuracy and relevance of your findings. By leveraging ChatGPT's capabilities and combining them with other research methods, you can develop more effective marketing and advertising strategies that resonate with your target audience and achieve your business objectives.

Content Creation for Marketing

ChatGPT can be a useful tool for generating ideas and creating content for marketing campaigns. Here's how you can use ChatGPT for content creation:

1. Brainstorm content ideas: Use ChatGPT to generate ideas for content that aligns with your marketing objectives and target audience. For example, "What are some blog post ideas for a beauty brand targeting millennials?" or "What are some video content ideas for a new line of eco-friendly home products?"
2. Develop brand messaging: Ask ChatGPT to help you develop clear and compelling brand messaging that resonates with your target audience. For example, "What are some key value propositions for a sustainable fashion brand?" or "What are some effective taglines for a new line of organic baby products?"
3. Create social media posts: Use ChatGPT to help you create engaging and shareable social media posts, such as captions, hashtags, or image descriptions. For example, "What are some creative Instagram caption ideas for a new line of workout equipment?" or "What are some effective hashtags to use for a new line of healthy snacks?"
4. Write blog posts or articles: ChatGPT can assist you in writing blog posts or articles by generating content ideas, outlining structures, or providing helpful information. For example, "What are some tips for writing an effective product review blog post?" or "What are the benefits of using natural ingredients in skincare products?"
5. Create email newsletters: Use ChatGPT to generate ideas for email newsletter content that resonates with your audience and prompts engagement. For example, "What are some engaging subject lines for a weekly newsletter for a tech startup?" or "What are some effective calls-to-action for an e-commerce store's holiday sale newsletter?"
6. Design infographics or visuals: ChatGPT can help you design visually appealing infographics or graphics that convey your message and resonate with your audience. For example, "What are some key statistics to include in an infographic about the benefits of meditation?" or "What are

some creative designs for a Facebook ad promoting a new line of activewear?"

Remember that ChatGPT should not be your only source of content creation ideas or guidance. It is important to cross-reference the insights and recommendations generated by ChatGPT with other sources, such as customer feedback, competitor analysis, or marketing experts, to ensure the accuracy and relevance of your findings. By leveraging ChatGPT's capabilities and combining them with other research methods, you can create more engaging and effective marketing content that resonates with your target audience and achieves your business objectives.

Customer Engagement and Interaction

ChatGPT can be a helpful tool for improving customer engagement and interaction by providing insights, suggestions, and personalized responses. Here's how you can use ChatGPT for customer engagement and interaction:

1. Personalize customer interactions: Use ChatGPT to personalize customer interactions by providing relevant information, suggestions, or responses based on their preferences, behavior, or history. For example, "What are some personalized recommendations for a customer who recently purchased a yoga mat?" or "What are some customized responses to a customer who is interested in eco-friendly products?"
2. Provide helpful information: ChatGPT can help you provide helpful information to customers by generating answers to their questions or inquiries, such as product specifications, delivery times, or return policies. For example, "What are the dimensions of a new line of furniture products?" or "What is the process for returning a defective product?"
3. Offer support and assistance: Ask ChatGPT to assist you in offering support and assistance to customers, such as troubleshooting technical issues, resolving complaints, or providing guidance on product usage. For example, "What are some steps to troubleshoot a customer's issue with a mobile app?" or "What are some tips for resolving a customer's complaint about a delayed delivery?"
4. Generate personalized recommendations: Use ChatGPT to generate personalized recommendations for customers based on their preferences, behavior, or purchase history. For example, "What are some product recommendations for a customer who has recently shown interest in organic skincare products?" or "What are some related products to recommend to a customer who has recently purchased a new laptop?"
5. Analyze customer feedback: ChatGPT can help you analyze customer feedback, such as reviews, ratings, or social media comments, to identify areas for improvement or opportunities to

enhance customer engagement and interaction.
6. Automate customer interactions: ChatGPT can assist you in automating customer interactions, such as chatbots, email responses, or personalized product recommendations, to improve efficiency and scalability while maintaining a high level of customer satisfaction.

Remember that ChatGPT should not replace human interaction and empathy in customer engagement and interaction. It is important to balance the use of ChatGPT with human support and expertise to provide personalized and effective customer service. By leveraging ChatGPT's capabilities and combining them with other customer service tools and strategies, you can improve customer engagement and interaction and build long-lasting customer relationships that benefit your business.

Chapter 6:
Customer Service Automation

Customer service automation using ChatGPT can be a useful tool for improving efficiency, scalability, and consistency in customer service while reducing costs and errors. Here's how you can use ChatGPT for customer service automation:

1. Develop chatbots: Chatbots can assist customers with common inquiries, such as product information, order status, or returns, using natural language processing and ChatGPT's capabilities. You can develop chatbots for different channels, such as website chat, social media messaging, or mobile apps, to improve accessibility and convenience for customers.
2. Automate email responses: Use ChatGPT to automate email responses to customers' inquiries, such as order confirmations, shipping notifications, or FAQs, to improve response time and reduce manual efforts. ChatGPT can generate customized responses based on customer data and previous interactions to provide a personalized experience.
3. Provide self-service options: ChatGPT can help you develop self-service options for customers, such as knowledge base articles, FAQs, or tutorials, to empower them to find answers to their questions and resolve their issues independently. This can improve efficiency and reduce the workload on your customer service team.
4. Automate personalized recommendations: Use ChatGPT to generate personalized product recommendations for customers based on their behavior, preferences, and purchase history. This can improve upselling and cross-selling opportunities and enhance the customer experience.
5. Analyze customer data: ChatGPT can assist you in analyzing customer data, such as purchase history, feedback, or behavior, to identify patterns, trends, and opportunities for improving customer service and engagement. This can help you optimize your customer service automation strategy and prioritize areas for improvement.

Remember that while ChatGPT can help automate customer service, it is essential to balance automation with human support and empathy to provide a personalized and satisfactory customer experience. Chatbots and automated responses should be designed to provide value and convenience for customers while maintaining a high level of quality and accuracy. It is also important to regularly evaluate and optimize your customer service automation strategy based on customer feedback and data analysis to ensure its effectiveness and relevance.

Role of ChatGPT in Customer Support

ChatGPT can play a significant role in improving customer support by providing efficient, personalized, and effective assistance to customers. Here's how you can leverage ChatGPT for customer support:

1. Automated FAQs: ChatGPT can assist you in developing automated Frequently Asked Questions (FAQs) for your customers to provide quick and easy access to information. This can help reduce the workload of your customer support team and enable customers to get the information they need quickly.
2. Personalized recommendations: Use ChatGPT to generate personalized recommendations for your customers based on their purchase history, preferences, and behavior. This can help customers discover new products they might be interested in and enhance their overall experience.
3. Natural language processing: ChatGPT's natural language processing (NLP) capabilities can enable your customers to interact with your support team in a conversational and natural way. Customers can ask questions in their own words, and ChatGPT can provide accurate and relevant responses in real-time.
4. Social media monitoring: ChatGPT can help you monitor social media channels for customer inquiries and complaints. By using ChatGPT to analyze customer feedback, you can identify patterns, trends, and opportunities for improving your customer support.
5. Automated responses: Use ChatGPT to automate responses to customer inquiries, such as order confirmations, shipping notifications, or FAQs. This can help improve response time and reduce the workload on your customer support team.
6. Chatbots: Chatbots can assist customers with common inquiries, such as product information, order status, or returns, using natural language processing and ChatGPT's capabilities. You can develop chatbots for different channels, such as website chat, social media messaging, or mobile apps, to improve accessibility and convenience for customers.

Remember that while ChatGPT can improve customer support, it should not replace human interaction and empathy. Chatbots and automated responses should be designed to provide value and convenience for customers while maintaining a high level of quality and accuracy. It is also important to regularly evaluate and optimize your ChatGPT-powered customer support strategy based on customer feedback and data analysis to ensure its effectiveness and relevance.

Setting Up Automated Customer Service

Here are the steps to setting up an automated customer service system using ChatGPT:

1. Define your customer service goals and objectives: Determine what you want to achieve with your customer service automation. Define your target metrics, such as response time, customer satisfaction, and resolution rate, and identify the areas of customer support that could benefit from automation.
2. Choose your customer service channels: Identify the channels your customers use to interact with your business, such as email, chat, social media, or phone, and choose the channels you want to automate.
3. Develop your customer service automation strategy: Develop a customer service automation strategy that aligns with your business objectives and customer needs. Determine the type of customer inquiries that can be automated, such as FAQs, order tracking, or returns, and the type of responses that can be generated by ChatGPT.
4. Build your ChatGPT models: Use a ChatGPT platform, such as GPT-3, to build your ChatGPT models. Train your models on your existing customer support data, including frequently asked questions, support tickets, and customer feedback. Use ChatGPT to generate personalized responses based on customer data and previous interactions to provide a customized experience.
5. Implement your customer service automation: Implement your customer service automation across your chosen channels, such as email auto-responses, chatbots, or social media monitoring. Integrate ChatGPT with your customer service platform or use an API to connect it with your website or mobile app.
6. Monitor and optimize your customer service automation: Regularly monitor your customer service automation to ensure that it is providing value and meeting your objectives. Use ChatGPT to analyze customer feedback and data to identify areas for improvement and optimize your automation strategy.

Remember that while ChatGPT can assist you in automating customer service, it should not replace human interaction and empathy entirely. Chatbots and automated responses should be designed to provide value and convenience for customers while maintaining a high level of quality and accuracy. It is also important to regularly evaluate and optimize your ChatGPT-powered customer service strategy based on customer feedback and data analysis to ensure its effectiveness and relevance.

Chapter 7:
ChatGPT for Product Development

ChatGPT can be a useful tool for product development by providing insights, suggestions, and feedback on various aspects of product design, development, and testing. Here are some ways to use ChatGPT for product development:

1. Idea generation: Use ChatGPT to generate new product ideas based on customer feedback, industry trends, or market research. For example, "What are some innovative product ideas for a new line of eco-friendly home products?" or "What are some unique features to include in a new mobile app for a fitness brand?"

2. Design feedback: Ask ChatGPT to provide feedback on product designs, such as packaging, user interface, or product features, to improve usability, functionality, and aesthetics. For example, "What are some improvements to make to a new line of sustainable clothing?" or "What are some design suggestions for a new line of smart home devices?"

3. User testing: ChatGPT can help you design and conduct user testing to gather feedback on product usability, functionality, and satisfaction. For example, "What are some questions to include in a user testing survey for a new mobile app?" or "What are some key metrics to track in a user testing session for a new line of beauty products?"

4. Product naming: Use ChatGPT to generate product names that are memorable, catchy, and relevant to your target audience. For example, "What are some effective product name suggestions for a new line of natural skincare products?" or "What are some creative product names for a new line of organic pet food?"

5. Marketing messaging: Ask ChatGPT to help you develop clear and compelling marketing messaging that resonates with your target audience and communicates the value of your product. For example, "What are some key value propositions for a new line of sustainable cleaning products?" or "What are some effective taglines for a new line of energy drinks?"

Remember that ChatGPT should not replace human expertise and creativity in product development. It is important to balance the use of ChatGPT with other research methods, such as customer feedback, competitor analysis, or expert insights, to ensure the accuracy, relevance, and creativity of your product development efforts. By leveraging ChatGPT's capabilities and combining them with other research methods and human expertise, you can develop innovative and effective products that meet your customers' needs and drive business growth.

Brainstorming Product Features with ChatGPT

ChatGPT can be a helpful tool for brainstorming product features by providing creative, diverse, and relevant ideas based on customer feedback, market trends, and industry insights. Here are some ways to use ChatGPT for brainstorming product features:

1. Define your product goals and objectives: Determine the purpose of your product and the value it should provide to your customers. Identify your target market, customer needs, and pain points, and use these as a basis for your brainstorming sessions.
2. Use open-ended questions: Ask ChatGPT open-ended questions to encourage a variety of responses and avoid limiting your creativity. For example, "What are some new features to add to a mobile app for a beauty brand?" or "What are some innovative product features to include in a new line of smart home devices?"
3. Combine different perspectives: Use ChatGPT to generate ideas from different perspectives, such as customer feedback, competitor analysis, or industry insights. Combine these perspectives to generate new and innovative ideas for product features.
4. Focus on user experience: Ask ChatGPT to provide ideas that improve the user experience, such as simplifying product usage, enhancing product functionality, or customizing product features. Focus on how the product can solve customer pain points and provide value to them.
5. Prioritize ideas: Use ChatGPT to prioritize your ideas based on their feasibility, impact, and relevance. Focus on the most impactful and feasible ideas and consider the resources and time required to implement them.

Remember that ChatGPT should not replace human creativity and expertise in product feature brainstorming. It is important to balance the use of ChatGPT with other research methods, such as customer feedback, competitor analysis, or expert insights, to ensure the accuracy, relevance, and creativity of your product development efforts. By leveraging ChatGPT's capabilities and combining them with other research methods and human expertise, you can develop innovative and effective product features that meet your customers' needs and drive business growth.

User Experience Design and ChatGPT

ChatGPT can be a useful tool for user experience (UX) design by providing insights, suggestions, and feedback on various aspects of UX, such as user research, user interface design, usability testing, and user engagement. Here are some ways to use ChatGPT for UX design:

1. User research: ChatGPT can help you conduct user research by generating survey questions,

interview questions, and user personas. For example, "What are some effective questions to include in a user research survey for a new mobile app?" or "What are some key characteristics to consider when developing user personas for a new line of eco-friendly products?"

2. User interface design: Ask ChatGPT to provide feedback on user interface design, such as color schemes, typography, layout, and navigation, to improve usability and aesthetics. For example, "What are some design suggestions for a new mobile app for a fitness brand?" or "What are some improvements to make to the user interface of a new line of smart home devices?"

3. Usability testing: ChatGPT can help you design and conduct usability testing to gather feedback on product usability, functionality, and satisfaction. For example, "What are some questions to include in a usability testing survey for a new line of beauty products?" or "What are some key metrics to track in a usability testing session for a new mobile app?"

4. User engagement: Use ChatGPT to generate ideas for improving user engagement, such as gamification, personalization, or social sharing. For example, "What are some ways to increase user engagement for a new line of sustainable clothing?" or "What are some effective techniques for incentivizing user engagement in a new mobile app?"

5. Accessibility: Ask ChatGPT to provide feedback on accessibility features, such as text-to-speech, closed captions, or screen readers, to improve the usability and accessibility of your product for users with disabilities.

Remember that ChatGPT should not replace human expertise and creativity in UX design. It is important to balance the use of ChatGPT with other research methods, such as user testing, heuristic evaluation, or expert review, to ensure the accuracy, relevance, and creativity of your UX design efforts. By leveraging ChatGPT's capabilities and combining them with other research methods and human expertise, you can develop user-friendly and engaging products that meet your customers' needs and drive business growth.

Chapter 8:
Improving Operational Efficiency

ChatGPT can be a useful tool for improving operational efficiency by providing insights, suggestions, and feedback on various aspects of operations, such as workflow optimization, process automation, resource allocation, and performance measurement. Here are some ways to use ChatGPT for improving operational efficiency:

1. Workflow optimization: Use ChatGPT to analyze your workflow and identify areas of inefficiency and bottlenecks. Ask ChatGPT to provide suggestions for optimizing workflow, such as reorganizing tasks, streamlining communication, or automating repetitive tasks.
2. Process automation: Ask ChatGPT to identify processes that can be automated to reduce manual effort and errors. For example, "What are some processes that can be automated in our customer service department?" or "What are some tools that can help automate our inventory management?"
3. Resource allocation: Use ChatGPT to analyze your resource allocation and identify areas of waste and inefficiency. Ask ChatGPT to provide suggestions for optimizing resource allocation, such as reallocating resources to high-priority tasks, outsourcing non-core tasks, or reducing unnecessary expenses.
4. Performance measurement: Ask ChatGPT to provide insights on performance measurement metrics to help you measure and improve operational efficiency. For example, "What are some key metrics to track in our manufacturing process?" or "What are some performance measurement tools to use in our project management?"
5. Decision making: Use ChatGPT to assist you in making informed decisions based on data analysis and industry insights. Ask ChatGPT to provide suggestions on decision-making frameworks, such as SWOT analysis, cost-benefit analysis, or risk analysis.

Remember that ChatGPT should not replace human expertise and judgment in improving operational efficiency. It is important to balance the use of ChatGPT with other research methods, such as data analysis, expert review, or stakeholder consultation, to ensure the accuracy, relevance, and creativity of your operational efficiency efforts. By leveraging ChatGPT's capabilities and combining them with other research methods and human expertise, you can improve operational efficiency and productivity while reducing costs and risks.

Process Automation with ChatGPT

ChatGPT can be a useful tool for process automation by providing insights, suggestions, and feedback on various aspects of automation, such as process identification, process design, process optimization, and process monitoring. Here are some ways to use ChatGPT for process automation:

1. Process identification: Use ChatGPT to identify processes that can be automated, such as repetitive tasks, data entry, or reporting. For example, "What are some processes in our customer service department that can be automated?" or "What are some tasks in our marketing department that can be automated?"

2. Process design: Ask ChatGPT to assist you in designing automated processes that are efficient, accurate, and scalable. For example, "What are some tools that can help automate our inventory management?" or "What are some best practices for designing automated workflows for our production process?"

3. Process optimization: Use ChatGPT to analyze your automated processes and identify areas of inefficiency, errors, or delays. Ask ChatGPT to provide suggestions for optimizing your automated processes, such as reducing manual intervention, improving data quality, or enhancing communication.

4. Process monitoring: Ask ChatGPT to help you monitor and measure the performance of your automated processes. Use ChatGPT to track key performance indicators (KPIs), such as cycle time, error rate, or throughput, and generate alerts when performance falls below a certain threshold.

5. Integration: Use ChatGPT to integrate your automated processes with other systems and tools, such as customer relationship management (CRM), enterprise resource planning (ERP), or business intelligence (BI) tools. Ask ChatGPT to provide suggestions for integration strategies and tools that are compatible with your existing systems and workflows.

Remember that ChatGPT should not replace human expertise and judgment in process automation. It is important to balance the use of ChatGPT with other research methods, such as process analysis, stakeholder consultation, or expert review, to ensure the accuracy, relevance, and creativity of your process automation efforts. By leveraging ChatGPT's capabilities and combining them with other research methods and human expertise, you can automate your processes to reduce manual effort, errors, and costs while improving efficiency, accuracy, and scalability.

Optimizing Workflows

ChatGPT can be a useful tool for optimizing workflows by providing insights, suggestions, and feedback on various aspects of workflow, such as task organization, communication, collaboration, and resource allocation. Here are some ways to use ChatGPT for optimizing workflows:

1. Task organization: Use ChatGPT to analyze your tasks and identify areas of inefficiency, duplication, or delay. Ask ChatGPT to provide suggestions for organizing tasks, such as prioritizing tasks, creating task templates, or categorizing tasks based on urgency and importance.
2. Communication: Ask ChatGPT to help you improve communication within your team by providing suggestions for better communication practices, such as using collaboration tools, setting up regular meetings, or creating communication protocols.
3. Collaboration: Use ChatGPT to help you improve collaboration within your team by providing suggestions for better collaboration practices, such as assigning roles and responsibilities, using collaboration tools, or creating a collaborative culture.
4. Resource allocation: Ask ChatGPT to assist you in optimizing your resource allocation by providing suggestions for better resource allocation practices, such as reallocating resources to high-priority tasks, outsourcing non-core tasks, or reducing unnecessary expenses.
5. Performance measurement: Use ChatGPT to assist you in measuring and improving workflow performance by providing suggestions for key performance indicators (KPIs) to track, such as cycle time, lead time, or task completion rate.

Remember that ChatGPT should not replace human expertise and judgment in optimizing workflows. It is important to balance the use of ChatGPT with other research methods, such as stakeholder consultation, expert review, or process analysis, to ensure the accuracy, relevance, and creativity of your workflow optimization efforts. By leveraging ChatGPT's capabilities and combining them with other research methods and human expertise, you can optimize your workflows to improve productivity, efficiency, and quality while reducing errors, delays, and costs.

Chapter 9:
ChatGPT for HR and Team Management

ChatGPT can be a useful tool for HR and team management by providing insights, suggestions, and feedback on various aspects of HR and team management, such as recruitment, employee engagement, performance management, and team building. Here are some ways to use ChatGPT for HR and team management:

1. Recruitment: Use ChatGPT to assist you in recruitment efforts by providing suggestions for job descriptions, candidate screening, and interview questions. For example, "What are some effective interview questions to ask when hiring for a project manager role?" or "What are some best practices for writing job descriptions for software engineers?"

2. Employee engagement: Ask ChatGPT to provide suggestions for improving employee engagement, such as employee recognition programs, team-building activities, or wellness initiatives. For example, "What are some effective ways to improve employee engagement for remote teams?" or "What are some fun team-building activities for a small startup?"

3. Performance management: Use ChatGPT to assist you in performance management efforts by providing suggestions for performance evaluation, feedback, and goal setting. For example, "What are some best practices for conducting a performance evaluation for a sales team?" or "What are some effective feedback techniques for remote teams?"

4. Team building: Ask ChatGPT to provide suggestions for team building activities that can improve collaboration, communication, and creativity. For example, "What are some team building activities that can help improve collaboration among different departments?" or "What are some creative team building ideas for a newly formed team?"

5. Professional development: Use ChatGPT to assist you in professional development efforts by providing suggestions for training programs, mentorship, and career planning. For example, "What are some effective training programs for software developers?" or "What are some best practices for mentorship programs in a startup environment?"

Remember that ChatGPT should not replace human expertise and judgment in HR and team management. It is important to balance the use of ChatGPT with other research methods, such as stakeholder consultation, expert review, or data analysis, to ensure the accuracy, relevance, and creativity of your HR and team management efforts. By leveraging ChatGPT's capabilities and combining them with other research methods and human expertise, you can improve HR and team management practices to enhance employee satisfaction, productivity, and retention.

Automated Employee Onboarding

ChatGPT can be a useful tool for automated employee onboarding by providing insights, suggestions, and feedback on various aspects of the onboarding process, such as paperwork, training, orientation, and team integration. Here are some ways to use ChatGPT for automated employee onboarding:

1. Paperwork: Use ChatGPT to assist you in automating paperwork by providing suggestions for digital forms, e-signature tools, and document management systems. For example, "What are some effective e-signature tools to use for onboarding paperwork?" or "What are some best practices for creating digital onboarding forms?"
2. Training: Ask ChatGPT to provide suggestions for automated training materials, such as video tutorials, online courses, or interactive quizzes. For example, "What are some effective video tutorials to include in a software developer onboarding program?" or "What are some best practices for creating online courses for a sales team?"
3. Orientation: Use ChatGPT to assist you in automating orientation sessions by providing suggestions for virtual onboarding sessions, video introductions, or gamified challenges. For example, "What are some effective virtual onboarding sessions to conduct for a remote team?" or "What are some creative ways to introduce new hires to the company culture?"
4. Team integration: Ask ChatGPT to provide suggestions for automated team integration practices, such as virtual meet-and-greets, buddy programs, or group projects. For example, "What are some effective buddy programs to implement in a startup environment?" or "What are some best practices for conducting virtual team-building activities?"

Remember that ChatGPT should not replace human interaction and support in employee onboarding. It is important to balance the use of ChatGPT with other human-led efforts, such as mentorship, coaching, and socialization, to ensure a comprehensive and personalized onboarding experience for new employees. By leveraging ChatGPT's capabilities and combining them with other human-led efforts, you can automate employee onboarding to reduce manual effort, errors, and costs while improving efficiency, accuracy, and scalability.

Team Communication and Coordination

ChatGPT can be a useful tool for team communication and coordination by providing insights, suggestions, and feedback on various aspects of team communication and coordination, such as tools and platforms, communication practices, conflict resolution, and goal alignment. Here are some ways to use ChatGPT for team communication and coordination:

1. Tools and platforms: Ask ChatGPT to provide suggestions for communication and collaboration tools and platforms that can improve team communication and coordination, such as project management software, team messaging apps, or video conferencing tools. For example, "What are some effective team messaging apps for a remote team?" or "What are some best practices for using project management software in a cross-functional team?"

2. Communication practices: Use ChatGPT to assist you in improving communication practices within your team by providing suggestions for better communication habits, such as clear and concise messaging, active listening, and timely feedback. For example, "What are some effective communication habits for a sales team?" or "What are some best practices for conducting remote team meetings?"

3. Conflict resolution: Ask ChatGPT to provide suggestions for conflict resolution strategies and techniques that can help resolve conflicts within your team, such as active listening, compromise, or mediation. For example, "What are some effective conflict resolution techniques for a cross-functional team?" or "What are some best practices for mediating conflicts in a startup environment?"

4. Goal alignment: Use ChatGPT to assist you in aligning team goals and objectives by providing suggestions for better goal-setting practices, such as SMART goals, OKRs, or KPIs. For example, "What are some effective goal-setting practices for a marketing team?" or "What are some best practices for measuring team performance in a software development team?"

Remember that ChatGPT should not replace human interaction and feedback in team communication and coordination. It is important to balance the use of ChatGPT with other human-led efforts, such as team building activities, coaching, and mentorship, to ensure effective team communication and coordination. By leveraging ChatGPT's capabilities and combining them with other human-led efforts, you can improve team communication and coordination to enhance team productivity, creativity, and satisfaction.

Chapter 10:
Looking Ahead: ChatGPT and the Future of Entrepreneurship

As artificial intelligence continues to develop and become more widely accessible, ChatGPT has the potential to become an increasingly valuable tool for entrepreneurs. With its ability to generate creative ideas, analyze data, and provide insights and suggestions, ChatGPT can help entrepreneurs to identify new business opportunities, optimize their operations, and improve customer engagement. Here are some potential applications of ChatGPT in the future of entrepreneurship:

1. Personalized customer experience: ChatGPT can help entrepreneurs to personalize their interactions with customers by providing tailored recommendations, responses, and solutions based on customer data and preferences.
2. Hyper-targeted marketing: ChatGPT can assist entrepreneurs in generating hyper-targeted marketing campaigns by analyzing customer behavior, interests, and demographics, and providing insights into the most effective messaging and channels for reaching their target audience.
3. Predictive analytics: ChatGPT can help entrepreneurs to forecast trends, anticipate demand, and identify potential risks and opportunities by analyzing data from various sources and providing insights into future market and customer behavior.
4. Automated business processes: ChatGPT can assist entrepreneurs in automating repetitive, time-consuming, and error-prone business processes, such as data entry, invoicing, and inventory management, by providing suggestions for tools and strategies that can streamline and optimize their operations.
5. Collaborative innovation: ChatGPT can help entrepreneurs to facilitate collaborative innovation by providing a platform for employees, customers, and partners to share ideas, feedback, and insights, and generate new solutions and products.

As with any technology, there are also potential risks and challenges associated with the use of ChatGPT in entrepreneurship, such as privacy concerns, bias and ethical issues, and the potential for overreliance on AI-generated insights. It is important for entrepreneurs to balance the use of ChatGPT with human expertise and judgment, and to stay informed about the latest developments and best practices in AI and entrepreneurship. By leveraging ChatGPT's capabilities and combining them with other research methods and human expertise, entrepreneurs can create innovative, sustainable, and socially responsible businesses that meet the needs and expectations of their customers and stakeholders.

Current Trends and Future Predictions

Current trends in entrepreneurship and AI suggest that the use of ChatGPT and other AI tools will continue to grow and expand in the coming years. Here are some current trends and future predictions related to ChatGPT and entrepreneurship:

1. Increased use of AI-powered customer service: With the rise of chatbots and virtual assistants, more companies are adopting AI-powered customer service to improve customer engagement and satisfaction. ChatGPT and other AI tools can help to enhance the capabilities and effectiveness of these systems, making them more personalized and responsive to customer needs.
2. Growing demand for AI-generated content: As content marketing becomes more important for businesses, the demand for AI-generated content, such as blog posts, social media updates, and product descriptions, is expected to increase. ChatGPT and other AI tools can help to automate and optimize the content creation process, making it more efficient and effective.
3. Greater use of AI in recruitment and HR: With the growing competition for talent, more companies are turning to AI-powered recruitment and HR tools to streamline their hiring process and improve employee engagement and retention. ChatGPT and other AI tools can assist in identifying and selecting the best candidates, providing personalized training and development, and facilitating team communication and collaboration.
4. Increasing focus on ethical and responsible AI: As AI becomes more pervasive in entrepreneurship and other areas of society, there is a growing awareness of the potential risks and challenges associated with its use. To address these concerns, there is likely to be an increasing focus on ethical and responsible AI practices, such as transparency, fairness, and privacy protection.
5. Advancements in natural language processing and chatbot technology: With advancements in natural language processing and chatbot technology, ChatGPT and other AI tools are expected to become more sophisticated and capable of understanding and responding to human language and behavior. This could lead to new applications in areas such as mental health, education, and personal finance.

Overall, the future of ChatGPT and entrepreneurship is likely to be characterized by increased integration and collaboration between AI and human expertise, as entrepreneurs continue to seek out innovative and effective ways to leverage AI technology to improve their businesses.

Ethical Considerations and Responsible Use of AI in Business

As AI technology continues to become more prevalent in business, it is important for entrepreneurs to consider the ethical implications of its use and adopt responsible practices to ensure that AI is used in a manner that is fair, transparent, and socially responsible. Here are some ethical considerations and responsible use practices for entrepreneurs using AI in business:

1. Avoiding bias and discrimination: One of the key ethical considerations when using AI in business is the potential for bias and discrimination. To avoid these issues, entrepreneurs should ensure that their AI systems are designed and trained using diverse and representative data sets, and regularly monitor their performance to identify and address any biases that may emerge.
2. Ensuring transparency and accountability: Entrepreneurs using AI should be transparent about how their systems are designed, what data they use, and how they make decisions. They should also provide clear channels for feedback and redress, so that customers and stakeholders can challenge decisions made by AI systems if necessary.
3. Protecting privacy and data security: Entrepreneurs using AI should take steps to protect the privacy and security of their customers' data, such as implementing robust encryption, access controls, and data retention policies. They should also comply with relevant data protection laws and regulations, and be transparent about how they collect, use, and share data.
4. Addressing the impact on employment: The use of AI in business has the potential to automate many tasks and functions that are currently performed by humans. Entrepreneurs should be mindful of the potential impact on employment, and take steps to reskill and retrain employees whose jobs may be at risk due to AI adoption.
5. Promoting ethical AI research and development: Entrepreneurs using AI should support and promote ethical AI research and development, such as initiatives focused on bias mitigation, explainability, and fairness. They should also engage with relevant stakeholders, including customers, employees, and regulators, to ensure that their AI systems are developed and used in a manner that is socially responsible and aligned with public values.

By adopting these ethical considerations and responsible use practices, entrepreneurs can ensure that their use of AI in business is ethical, fair, and aligned with societal values, while also maximizing the potential benefits of AI technology for their businesses and stakeholders.

The best PROMPT for the ADVERTISER:

Write in GPT Chat this prompt exactly as it is written below. Try changing the terms you find in the " " to get the work that works best for you. Remember that in case you are an advertiser this prompt is PERFECT for you, it is a very powerful tool make good use of it!

Prompt for Sales Page:

You are an advertiser and will act as such. You will write an "informational" themed sales page titled "The Best Bicycle You Could Ever Buy," for an audience of "male, over 50" aiming to convert the reader into a buyer. The Article will have the following structure: Title, introduction, 1 H2 sections, 1 H3 sections, .The H3 sections will be sub sections of the H2 sections. You will format the story in markdown. You will use a "professional" writing style, an "engaging" sentiment, and an "informal" communicative register.

Prompt for Market Research:

You are an analyst and you will act like one. I am an advertiser and I will use this information to then create an advertisement. Keeping in mind that I want to sell this "bicycle" product, do extensive market research with the goal of informing me of all the details that the buyer person might have to make writing my advertisement easier for me. The informative article should be as long and detailed as possible, it will have the following structure: Title, introduction, 1 H2 section, 1 H3 section .The H3 sections will be subsections of the H2 sections. You will format the article in markdown. You will use an "analytical" writing style, a "formal" feeling, and a "professional" communicative register.

Prompt for Newsletter:

You are a social media manager and you will act like one. Keeping in mind that I have a business based on "yoga," create a comprehensive newsletter publishing plan. This newsletter plan should be as long and detailed as possible and will serve to retain customers, it will have the following structure: Title, introduction, 1 H2 section, 1 H3 section. The H3 sections will be subsections of the H2 sections. You will format the article in markdown. You will use an "analytical" writing style, a "formal" tone, and a "professional" communicative register.

BOOK 3: "ChatGPT for Researchers: Assistance in Research and Data Analysis": Shows how ChatGPT can be used as a research and analysis tool.

Chapter 1:
Introduction to ChatGPT for Researchers

ChatGPT is an advanced natural language processing (NLP) tool that can be used by researchers to generate new insights, ideas, and hypotheses, as well as to analyze and interpret data. With its ability to process large amounts of text data, identify patterns and trends, and generate human-like responses, ChatGPT can assist researchers in a wide range of fields, including social sciences, natural sciences, engineering, and humanities. In this context, ChatGPT can be seen as a useful tool for assisting researchers in identifying new research areas, generating research questions, and analyzing research data.

Some potential applications of ChatGPT for researchers include:

1. Automated literature review: ChatGPT can assist researchers in conducting automated literature reviews, by providing summaries and analysis of existing research papers, identifying gaps in the literature, and suggesting new research directions.
2. Idea generation: ChatGPT can help researchers to generate new ideas and hypotheses by analyzing and synthesizing large amounts of data, identifying patterns and relationships, and proposing novel research questions and hypotheses.
3. Data analysis and interpretation: ChatGPT can assist researchers in analyzing and interpreting data, by providing insights and suggestions for data visualization, statistical analysis, and machine learning models.
4. Natural language processing: ChatGPT can be used to process and analyze text data, such as social media posts, news articles, and online reviews, to extract insights and trends, and identify patterns and relationships.

It is important to note that while ChatGPT can assist researchers in generating ideas and analyzing data, it should not replace human judgment and expertise. Researchers should use ChatGPT in combination with other research methods, such as literature review, interviews, surveys, and experiments, to ensure that their research is comprehensive, rigorous, and valid. Additionally, researchers should be aware of the limitations and potential biases of ChatGPT, and take steps to mitigate these risks, such as by validating ChatGPT-generated insights with other sources of data and expertise.

Understanding ChatGPT

ChatGPT is an artificial intelligence (AI) language model developed by OpenAI. It is based on a deep learning architecture called the Transformer, which allows it to process and generate natural language text with high accuracy and fluency. ChatGPT has been trained on a massive amount of text data, including books, articles, and web pages, and can generate responses that are often indistinguishable from those of a human.

ChatGPT works by processing a sequence of input tokens, such as words or sentences, and predicting the most likely sequence of output tokens based on the context of the input. It does this by using a combination of unsupervised and supervised learning techniques, including language modeling, attention mechanisms, and fine-tuning.

One of the key features of ChatGPT is its ability to generate coherent and contextually relevant responses to a wide range of prompts and questions. It can understand and respond to natural language text on a variety of topics, such as current events, science, technology, and entertainment, and can generate responses that are informative, creative, and engaging.

In addition to generating text, ChatGPT can also be used for a variety of NLP tasks, such as sentiment analysis, text classification, and language translation. It can assist in generating summaries and insights from large amounts of text data, such as news articles or social media posts, and can help to automate many aspects of the text analysis process.

Overall, ChatGPT represents a significant advance in the field of AI and NLP, and has the potential to transform many aspects of research, business, and society. By leveraging its capabilities in natural language processing and text generation, researchers can generate new insights and hypotheses, businesses can improve their customer engagement and operational efficiency, and society can benefit from new applications in education, healthcare, and entertainment.

The Role of AI in Research

The role of AI in research is becoming increasingly important, as AI technologies like ChatGPT can assist researchers in generating new insights, analyzing data, and improving research processes. Here are some ways that AI can be used in research:

1. Automated literature reviews: AI can be used to automate literature reviews, by analyzing large amounts of text data and identifying patterns and trends. This can save researchers significant time and effort in conducting manual literature reviews, and can help to identify research gaps

and new areas for investigation.

2. Data analysis and interpretation: AI can be used to analyze and interpret data, by identifying patterns, correlations, and relationships in large and complex data sets. This can help researchers to gain new insights and generate hypotheses, and can improve the accuracy and reliability of research findings.

3. Natural language processing: AI can be used to process and analyze text data, such as survey responses or social media posts, to identify sentiment, topics, and themes. This can help researchers to understand how people are thinking and feeling about a particular topic or issue, and can provide new insights into social trends and behaviors.

4. Predictive modeling: AI can be used to build predictive models, by identifying patterns and relationships in data and making predictions about future outcomes. This can help researchers to forecast trends and make informed decisions based on data-driven insights.

5. Collaborative research: AI can be used to facilitate collaborative research, by providing a platform for researchers to share data, ideas, and insights, and to collaborate on research projects in real-time.

It is important to note that while AI can be a powerful tool in research, it should not replace human judgment and expertise. Researchers should use AI in combination with other research methods and approaches, and should be aware of the potential biases and limitations of AI-generated insights. Additionally, researchers should ensure that their use of AI is ethical and responsible, and that they comply with relevant data protection and privacy laws and regulations.

Chapter 2:
Literature Review with ChatGPT

ChatGPT can be used to automate the process of conducting a literature review, which is an important aspect of any research project. Here are some ways that ChatGPT can assist with literature reviews:

1. Identifying key concepts and themes: ChatGPT can analyze large volumes of text data, such as academic papers, books, and other publications, to identify key concepts, themes, and topics. This can help to streamline the literature review process, and ensure that relevant literature is identified and included in the review.
2. Summarizing research findings: ChatGPT can generate summaries of research papers and publications, which can help to quickly identify relevant research findings and conclusions. This can save researchers time and effort, and provide a clear overview of the literature in a particular field.
3. Identifying research gaps: ChatGPT can help to identify research gaps and areas for further investigation by analyzing existing literature and identifying areas where further research is needed. This can help researchers to generate new research questions and hypotheses, and can ensure that their research is original and valuable.
4. Synthesizing information: ChatGPT can assist in synthesizing information from multiple sources, such as academic papers or other publications, to provide a more comprehensive and nuanced understanding of a particular topic. This can help researchers to identify patterns and relationships in the literature, and to gain new insights into their research area.
5. Providing recommendations: ChatGPT can provide recommendations for further reading and research, based on the topics and themes identified in the literature review. This can help researchers to identify new research directions, and to ensure that their research is up-to-date and informed by the latest literature.

It is important to note that while ChatGPT can assist with literature reviews, it should not replace human judgment and expertise. Researchers should use ChatGPT in combination with other research methods and approaches, and should be aware of the potential biases and limitations of ChatGPT-generated insights. Additionally, researchers should ensure that their use of ChatGPT is ethical and responsible, and that they comply with relevant data protection and privacy laws and regulations.

Summarizing Articles with ChatGPT

ChatGPT can be used to summarize articles and other written content, which can be a useful tool for researchers who need to quickly understand the key points and findings of a particular publication. Here are some ways that ChatGPT can assist with article summarization:

1. Identifying key topics and themes: ChatGPT can analyze the text of an article and identify the key topics and themes that it covers. This can help to provide a clear overview of the content, and highlight the most important aspects of the article.
2. Extracting key sentences and phrases: ChatGPT can identify the most important sentences and phrases in an article, and use them to generate a summary that accurately captures the main points and findings of the article.
3. Providing a concise summary: ChatGPT can generate a concise summary of an article, which can be useful for researchers who need to quickly understand the key points and findings. This can save time and effort, and provide a clear overview of the content.
4. Improving readability: ChatGPT can help to improve the readability of an article summary, by removing redundant or unnecessary information and focusing on the most important points. This can make the summary more accessible and easier to understand.
5. Generating multiple summaries: ChatGPT can be used to generate multiple summaries of the same article, which can provide different perspectives and insights. This can be useful for researchers who are looking for a more comprehensive understanding of the article.

It is important to note that while ChatGPT can assist with article summarization, it should not replace human judgment and expertise. Researchers should use ChatGPT-generated summaries in combination with other research methods and approaches, and should be aware of the potential biases and limitations of ChatGPT-generated insights. Additionally, researchers should ensure that their use of ChatGPT is ethical and responsible, and that they comply with relevant data protection and privacy laws and regulations.

Using ChatGPT for Comprehensive Literature Searches

ChatGPT can be used to assist with comprehensive literature searches, which are an important aspect of any research project. Here are some ways that ChatGPT can assist with literature searches:

1. Identifying relevant keywords and phrases: ChatGPT can analyze a research question or topic and identify relevant keywords and phrases that can be used in a literature search. This can help to ensure that the search is comprehensive and includes all relevant literature.

2. Conducting searches across multiple databases: ChatGPT can be used to search multiple academic databases, such as PubMed or Scopus, to identify relevant research papers and publications. This can save researchers time and effort, and ensure that the literature search is comprehensive and up-to-date.
3. Filtering search results: ChatGPT can assist in filtering search results based on various criteria, such as publication date, author, or journal. This can help to ensure that only relevant papers are included in the literature review.
4. Identifying gaps in the literature: ChatGPT can assist in identifying gaps in the literature by analyzing the search results and identifying areas where further research is needed. This can help researchers to generate new research questions and hypotheses, and ensure that their research is original and valuable.
5. Providing recommendations: ChatGPT can provide recommendations for further reading and research based on the search results, which can help researchers to identify new research directions and ensure that their research is up-to-date and informed by the latest literature.

It is important to note that while ChatGPT can assist with literature searches, it should not replace human judgment and expertise. Researchers should use ChatGPT in combination with other research methods and approaches, and should be aware of the potential biases and limitations of ChatGPT-generated insights. Additionally, researchers should ensure that their use of ChatGPT is ethical and responsible, and that they comply with relevant data protection and privacy laws and regulations.

Chapter 3:
Hypothesis Generation and Validation

ChatGPT can be used to assist with hypothesis generation and validation, which are critical steps in any research project. Here are some ways that ChatGPT can assist with hypothesis generation and validation:

1. Generating new research questions: ChatGPT can be used to generate new research questions based on a topic or area of interest. Researchers can input a topic or research question into ChatGPT, and the model will generate a list of related questions that may be useful for further investigation.
2. Identifying potential variables: ChatGPT can assist in identifying potential variables that may be relevant to a particular research question or hypothesis. Researchers can input a research question into ChatGPT, and the model will generate a list of related variables that may be useful for further investigation.
3. Testing hypotheses: ChatGPT can be used to test hypotheses by analyzing data and identifying patterns and relationships. Researchers can input data into ChatGPT, and the model will analyze the data and generate insights and predictions based on the input.
4. Identifying research gaps: ChatGPT can assist in identifying research gaps by analyzing existing literature and identifying areas where further research is needed. This can help researchers to generate new research questions and hypotheses, and can ensure that their research is original and valuable.
5. Providing recommendations: ChatGPT can provide recommendations for further research based on the input data and research questions. This can help researchers to identify new research directions and ensure that their research is up-to-date and informed by the latest literature.

It is important to note that while ChatGPT can assist with hypothesis generation and validation, it should not replace human judgment and expertise. Researchers should use ChatGPT in combination with other research methods and approaches, and should be aware of the potential biases and limitations of ChatGPT-generated insights. Additionally, researchers should ensure that their use of ChatGPT is ethical and responsible, and that they comply with relevant data protection and privacy laws and regulations.

Leveraging ChatGPT for Innovative Hypotheses

ChatGPT can be a valuable tool for generating innovative hypotheses in research. Here are some ways that ChatGPT can be leveraged for innovative hypothesis generation:

1. Inputting open-ended questions: ChatGPT is particularly effective at generating innovative hypotheses when researchers input open-ended questions, rather than closed-ended questions. Open-ended questions allow ChatGPT to generate more creative and diverse responses, which can lead to new and innovative hypotheses.
2. Exploring related topics: Researchers can use ChatGPT to explore related topics to their research area, which can lead to the generation of innovative hypotheses. By inputting related keywords and phrases into ChatGPT, researchers can explore new areas of inquiry and discover innovative hypotheses.
3. Combining multiple sources of data: ChatGPT can be used to combine multiple sources of data, such as survey responses, social media posts, and academic literature, to generate innovative hypotheses. By inputting data from multiple sources into ChatGPT, researchers can identify patterns and relationships that may not be apparent from a single data source.
4. Collaborating with others: Researchers can collaborate with others using ChatGPT, which can lead to the generation of innovative hypotheses. By inputting data and ideas from multiple researchers into ChatGPT, innovative hypotheses can be generated that draw on diverse perspectives and areas of expertise.
5. Using generative prompts: Researchers can use ChatGPT to generate prompts for innovative hypothesis generation. By inputting a topic or research question into ChatGPT, researchers can generate a list of prompts that can be used to generate innovative hypotheses.

It is important to note that while ChatGPT can assist with hypothesis generation, it should not replace human judgment and expertise. Researchers should use ChatGPT in combination with other research methods and approaches, and should be aware of the potential biases and limitations of ChatGPT-generated insights. Additionally, researchers should ensure that their use of ChatGPT is ethical and responsible, and that they comply with relevant data protection and privacy laws and regulations.

Validating Hypotheses Using AI

ChatGPT can be used to assist with the validation of hypotheses in research. Here are some ways that ChatGPT can be leveraged for hypothesis validation:

1. Analyzing data: ChatGPT can be used to analyze data and identify patterns and relationships that can be used to validate hypotheses. Researchers can input data into ChatGPT, and the model will analyze the data and generate insights and predictions based on the input.
2. Conducting simulations: ChatGPT can be used to conduct simulations to test the validity of

hypotheses. By inputting data and hypotheses into ChatGPT, researchers can simulate different scenarios and test the validity of their hypotheses.

3. Identifying outliers: ChatGPT can be used to identify outliers in data that may be relevant to the validation of hypotheses. By inputting data into ChatGPT, researchers can identify patterns and relationships that may not be apparent from a simple analysis of the data.

4. Comparing hypotheses: ChatGPT can be used to compare different hypotheses and identify the most likely explanation for the data. By inputting different hypotheses into ChatGPT, researchers can compare them and determine which one is most likely to be true.

5. Providing recommendations: ChatGPT can provide recommendations for further research based on the input data and hypotheses. This can help researchers to identify new research directions and ensure that their research is up-to-date and informed by the latest literature.

It is important to note that while ChatGPT can assist with hypothesis validation, it should not replace human judgment and expertise. Researchers should use ChatGPT in combination with other research methods and approaches, and should be aware of the potential biases and limitations of ChatGPT-generated insights. Additionally, researchers should ensure that their use of ChatGPT is ethical and responsible, and that they comply with relevant data protection and privacy laws and regulations.

Chapter 4:
Research Design and Methodology

ChatGPT can be used to assist with research design and methodology, which are critical steps in any research project. Here are some ways that ChatGPT can be leveraged for research design and methodology:

1. Generating research questions: ChatGPT can be used to generate research questions that are innovative and original. By inputting related keywords and phrases into ChatGPT, researchers can explore new areas of inquiry and generate research questions that are grounded in the latest literature.
2. Designing experiments: ChatGPT can be used to design experiments that are robust and reliable. By inputting data and hypotheses into ChatGPT, researchers can simulate different scenarios and test the validity of their hypotheses, which can help to ensure that the experimental design is appropriate.
3. Developing survey questions: ChatGPT can be used to develop survey questions that are clear and easy to understand. By inputting questions into ChatGPT, researchers can generate multiple versions of the question that are designed to be more effective and informative.
4. Analyzing data: ChatGPT can be used to analyze data and identify patterns and relationships that are relevant to the research questions. By inputting data into ChatGPT, researchers can generate insights and predictions that can help to refine the research methodology.
5. Providing recommendations: ChatGPT can provide recommendations for further research based on the input data and hypotheses. This can help researchers to identify new research directions and ensure that their research is up-to-date and informed by the latest literature.

It is important to note that while ChatGPT can assist with research design and methodology, it should not replace human judgment and expertise. Researchers should use ChatGPT in combination with other research methods and approaches, and should be aware of the potential biases and limitations of ChatGPT-generated insights. Additionally, researchers should ensure that their use of ChatGPT is ethical and responsible, and that they comply with relevant data protection and privacy laws and regulations.

Using ChatGPT for Research Design Ideas

ChatGPT can be used to assist with research design ideas, which are critical for designing rigorous and effective research studies. Here are some ways that ChatGPT can be leveraged for research design ideas:

1. Inputting research questions: ChatGPT can be used to generate research questions that can be used to guide research design. By inputting related keywords and phrases into ChatGPT, researchers can generate a list of research questions that are grounded in the latest literature.

2. Identifying potential variables: ChatGPT can assist in identifying potential variables that may be relevant to a particular research question or hypothesis. Researchers can input a research question into ChatGPT, and the model will generate a list of related variables that may be useful for further investigation.

3. Designing experiments: ChatGPT can be used to design experiments that are robust and reliable. By inputting data and hypotheses into ChatGPT, researchers can simulate different scenarios and test the validity of their hypotheses, which can help to ensure that the experimental design is appropriate.

4. Generating survey questions: ChatGPT can be used to generate survey questions that are clear and easy to understand. By inputting questions into ChatGPT, researchers can generate multiple versions of the question that are designed to be more effective and informative.

5. Providing recommendations: ChatGPT can provide recommendations for further research based on the input data and research questions. This can help researchers to identify new research directions and ensure that their research is up-to-date and informed by the latest literature.

It is important to note that while ChatGPT can assist with research design ideas, it should not replace human judgment and expertise. Researchers should use ChatGPT in combination with other research methods and approaches, and should be aware of the potential biases and limitations of ChatGPT-generated insights. Additionally, researchers should ensure that their use of ChatGPT is ethical and responsible, and that they comply with relevant data protection and privacy laws and regulations.

Evaluating Methodologies with ChatGPT

ChatGPT can be used to assist with evaluating methodologies in research. Here are some ways that ChatGPT can be leveraged for evaluating methodologies:

1. Identifying potential biases: ChatGPT can assist in identifying potential biases in research methodology by analyzing the data and identifying patterns and relationships. By inputting data into ChatGPT, researchers can identify potential biases in the methodology and make adjustments to improve the validity of the research.

2. Testing validity: ChatGPT can be used to test the validity of research methodology by analyzing the data and identifying patterns and relationships. By inputting data into ChatGPT, researchers

can simulate different scenarios and test the validity of the methodology, which can help to ensure that the research is reliable.

3. Evaluating research design: ChatGPT can be used to evaluate the research design and make adjustments to improve the validity of the research. By inputting data and hypotheses into ChatGPT, researchers can simulate different scenarios and evaluate the effectiveness of the research design.

4. Providing recommendations: ChatGPT can provide recommendations for improving the research methodology based on the input data and research questions. This can help researchers to identify areas for improvement and ensure that their research is up-to-date and informed by the latest literature.

5. Analyzing data: ChatGPT can be used to analyze data and identify patterns and relationships that are relevant to the research questions. By inputting data into ChatGPT, researchers can generate insights and predictions that can help to refine the research methodology.

It is important to note that while ChatGPT can assist with evaluating methodologies, it should not replace human judgment and expertise. Researchers should use ChatGPT in combination with other research methods and approaches, and should be aware of the potential biases and limitations of ChatGPT-generated insights. Additionally, researchers should ensure that their use of ChatGPT is ethical and responsible, and that they comply with relevant data protection and privacy laws and regulations.

Chapter 5:
ChatGPT for Data Collection

ChatGPT can be used to assist with data collection in research. Here are some ways that ChatGPT can be leveraged for data collection:

1. Generating survey questions: ChatGPT can be used to generate survey questions that are clear and easy to understand. By inputting questions into ChatGPT, researchers can generate multiple versions of the question that are designed to be more effective and informative.
2. Identifying potential variables: ChatGPT can assist in identifying potential variables that may be relevant to a particular research question or hypothesis. Researchers can input a research question into ChatGPT, and the model will generate a list of related variables that may be useful for further investigation.
3. Collecting and organizing data: ChatGPT can be used to collect and organize data, particularly unstructured data such as social media posts or customer feedback. By inputting data into ChatGPT, researchers can organize the data and identify patterns and relationships that may be relevant to the research questions.
4. Conducting sentiment analysis: ChatGPT can be used to conduct sentiment analysis on data, such as social media posts or customer feedback. By inputting data into ChatGPT, researchers can analyze the sentiment of the data and identify patterns and trends in customer sentiment.
5. Providing recommendations: ChatGPT can provide recommendations for further data collection based on the input data and research questions. This can help researchers to identify new areas of inquiry and ensure that their research is up-to-date and informed by the latest data.

It is important to note that while ChatGPT can assist with data collection, it should not replace human judgment and expertise. Researchers should use ChatGPT in combination with other data collection methods and approaches, and should be aware of the potential biases and limitations of ChatGPT-generated insights. Additionally, researchers should ensure that their use of ChatGPT is ethical and responsible, and that they comply with relevant data protection and privacy laws and regulations.

Designing Surveys and Questionnaires

ChatGPT can be used to assist with designing surveys and questionnaires in research. Here are some ways that ChatGPT can be leveraged for survey and questionnaire design:

1. Generating survey questions: ChatGPT can be used to generate survey questions that are clear and easy to understand. By inputting keywords and phrases related to the research topic into ChatGPT, researchers can generate multiple versions of the question that are designed to be more effective and informative.
2. Analyzing existing surveys: ChatGPT can be used to analyze existing surveys and identify areas for improvement. By inputting an existing survey into ChatGPT, researchers can identify patterns and relationships in the responses and determine which questions may be redundant or unclear.
3. Tailoring surveys to specific populations: ChatGPT can be used to tailor surveys to specific populations by generating questions that are relevant and appropriate for the population being surveyed. By inputting keywords related to the population into ChatGPT, researchers can generate questions that are more likely to be understood and answered accurately by that population.
4. Conducting pilot studies: ChatGPT can be used to conduct pilot studies of surveys and questionnaires. By inputting data from pilot studies into ChatGPT, researchers can simulate different scenarios and test the effectiveness of the survey questions.
5. Providing recommendations: ChatGPT can provide recommendations for further survey design based on the input data and research questions. This can help researchers to identify new areas of inquiry and ensure that their surveys are up-to-date and informed by the latest literature.

It is important to note that while ChatGPT can assist with survey and questionnaire design, it should not replace human judgment and expertise. Researchers should use ChatGPT in combination with other survey design methods and approaches, and should be aware of the potential biases and limitations of ChatGPT-generated insights. Additionally, researchers should ensure that their use of ChatGPT is ethical and responsible, and that they comply with relevant data protection and privacy laws and regulations.

Conducting Interviews with ChatGPT

ChatGPT can be used to assist with conducting interviews in research. Here are some ways that ChatGPT can be leveraged for conducting interviews:

1. Generating interview questions: ChatGPT can be used to generate interview questions that are relevant and insightful. By inputting keywords related to the research topic into ChatGPT, researchers can generate multiple versions of the question that are designed to be more effective and informative.
2. Analyzing interview data: ChatGPT can be used to analyze interview data and identify patterns

and relationships in the responses. By inputting interview transcripts into ChatGPT, researchers can identify common themes and topics that may be relevant to the research questions.

3. Tailoring interviews to specific populations: ChatGPT can be used to tailor interviews to specific populations by generating questions that are relevant and appropriate for the population being interviewed. By inputting keywords related to the population into ChatGPT, researchers can generate questions that are more likely to be understood and answered accurately by that population.

4. Conducting pilot studies: ChatGPT can be used to conduct pilot studies of interview questions. By inputting interview data from pilot studies into ChatGPT, researchers can simulate different scenarios and test the effectiveness of the interview questions.

5. Providing recommendations: ChatGPT can provide recommendations for further interview design based on the input data and research questions. This can help researchers to identify new areas of inquiry and ensure that their interviews are up-to-date and informed by the latest literature.

It is important to note that while ChatGPT can assist with interview design and analysis, it should not replace human judgment and expertise. Researchers should use ChatGPT in combination with other interview methods and approaches, and should be aware of the potential biases and limitations of ChatGPT-generated insights. Additionally, researchers should ensure that their use of ChatGPT is ethical and responsible, and that they comply with relevant data protection and privacy laws and regulations.

Chapter 6:
Data Analysis with ChatGPT

ChatGPT can be used to assist with data analysis in research. Here are some ways that ChatGPT can be leveraged for data analysis:

1. Identifying patterns and relationships: ChatGPT can be used to identify patterns and relationships in large datasets. By inputting data into ChatGPT, researchers can generate insights and predictions that can help to refine the research methodology.
2. Conducting sentiment analysis: ChatGPT can be used to conduct sentiment analysis on data, such as social media posts or customer feedback. By inputting data into ChatGPT, researchers can analyze the sentiment of the data and identify patterns and trends in customer sentiment.
3. Generating hypotheses: ChatGPT can be used to generate hypotheses based on patterns and relationships in the data. By inputting data into ChatGPT, researchers can generate hypotheses that can be further tested and validated.
4. Conducting simulations: ChatGPT can be used to conduct simulations of different scenarios based on the input data. By inputting data into ChatGPT, researchers can simulate different scenarios and test the effectiveness of different strategies.
5. Providing recommendations: ChatGPT can provide recommendations for further data analysis based on the input data and research questions. This can help researchers to identify new areas of inquiry and ensure that their research is up-to-date and informed by the latest data.

It is important to note that while ChatGPT can assist with data analysis, it should not replace human judgment and expertise. Researchers should use ChatGPT in combination with other data analysis methods and approaches, and should be aware of the potential biases and limitations of ChatGPT-generated insights. Additionally, researchers should ensure that their use of ChatGPT is ethical and responsible, and that they comply with relevant data protection and privacy laws and regulations.

Qualitative Data Analysis

ChatGPT can be used to assist with qualitative data analysis in research. Here are some ways that ChatGPT can be leveraged for qualitative data analysis:

1. Identifying themes: ChatGPT can be used to identify themes in qualitative data such as interview transcripts, focus group discussions, and open-ended survey responses. By inputting the data

into ChatGPT, researchers can identify common words or phrases and determine themes that are emerging from the data.
2. Conducting content analysis: ChatGPT can be used to conduct content analysis on qualitative data. By inputting the data into ChatGPT, researchers can identify patterns and relationships in the data that may be relevant to the research questions.
3. Providing context: ChatGPT can be used to provide context for qualitative data. By inputting the data into ChatGPT, researchers can generate explanations or background information that can help to better understand the data and the context in which it was collected.
4. Analyzing sentiment: ChatGPT can be used to analyze the sentiment of qualitative data such as social media posts or customer feedback. By inputting the data into ChatGPT, researchers can analyze the sentiment of the data and identify patterns and trends in customer sentiment.
5. Generating hypotheses: ChatGPT can be used to generate hypotheses based on themes or patterns in the qualitative data. By inputting the data into ChatGPT, researchers can generate hypotheses that can be further tested and validated.

It is important to note that while ChatGPT can assist with qualitative data analysis, it should not replace human judgment and expertise. Researchers should use ChatGPT in combination with other qualitative data analysis methods and approaches, and should be aware of the potential biases and limitations of ChatGPT-generated insights. Additionally, researchers should ensure that their use of ChatGPT is ethical and responsible, and that they comply with relevant data protection and privacy laws and regulations.

Quantitative Data Analysis

ChatGPT can be used to assist with quantitative data analysis in research. Here are some ways that ChatGPT can be leveraged for quantitative data analysis:

1. Descriptive statistics: ChatGPT can be used to generate descriptive statistics for quantitative data, such as means, standard deviations, and frequencies. By inputting the data into ChatGPT, researchers can generate summary statistics that can provide a basic understanding of the data.
2. Inferential statistics: ChatGPT can be used to conduct inferential statistics on quantitative data, such as t-tests or ANOVA. By inputting the data into ChatGPT, researchers can analyze the relationships between variables and identify significant differences or correlations.
3. Data visualization: ChatGPT can be used to generate data visualizations for quantitative data, such as histograms or scatterplots. By inputting the data into ChatGPT, researchers can create visual representations of the data that can help to identify patterns and trends.
4. Time series analysis: ChatGPT can be used to conduct time series analysis on quantitative data,

such as stock prices or website traffic. By inputting the data into ChatGPT, researchers can identify patterns over time and make predictions about future trends.

5. Machine learning: ChatGPT can be used to conduct machine learning on quantitative data, such as clustering or classification. By inputting the data into ChatGPT, researchers can identify patterns and relationships in the data that may be difficult to detect through traditional statistical analysis.

It is important to note that while ChatGPT can assist with quantitative data analysis, it should not replace human judgment and expertise. Researchers should use ChatGPT in combination with other quantitative data analysis methods and approaches, and should be aware of the potential biases and limitations of ChatGPT-generated insights. Additionally, researchers should ensure that their use of ChatGPT is ethical and responsible, and that they comply with relevant data protection and privacy laws and regulations.

Chapter 7:
ChatGPT for Writing Research Papers

ChatGPT can be used to assist with writing research papers in several ways:

1. Generating topic ideas: ChatGPT can be used to generate topic ideas for research papers based on keywords related to the research area. By inputting keywords into ChatGPT, researchers can generate a list of potential topics and select the most relevant one for their research paper.
2. Conducting literature reviews: ChatGPT can be used to assist with conducting literature reviews for research papers. By inputting keywords related to the research area, ChatGPT can generate summaries of relevant research articles and identify common themes and research gaps.
3. Writing outlines: ChatGPT can be used to generate outlines for research papers based on the research questions and objectives. By inputting the research questions and objectives into ChatGPT, researchers can generate an outline that includes the main sections of the research paper.
4. Writing introductions and conclusions: ChatGPT can be used to assist with writing introductions and conclusions for research papers. By inputting the research questions and objectives into ChatGPT, researchers can generate an introduction that provides background information and a conclusion that summarizes the main findings of the research paper.
5. Editing and proofreading: ChatGPT can be used to assist with editing and proofreading research papers. By inputting the research paper into ChatGPT, researchers can identify grammatical errors, spelling mistakes, and other issues that need to be corrected.

It is important to note that while ChatGPT can assist with writing research papers, it should not replace human judgment and expertise. Researchers should use ChatGPT in combination with other writing methods and approaches, and should be aware of the potential biases and limitations of ChatGPT-generated insights. Additionally, researchers should ensure that their use of ChatGPT is ethical and responsible, and that they comply with relevant data protection and privacy laws and regulations.

Structuring Your Paper with ChatGPT

ChatGPT can be used to assist with structuring research papers by generating outlines and organizing ideas. Here are some ways that ChatGPT can be leveraged to structure research papers:

1. Generating outlines: ChatGPT can be used to generate outlines for research papers based on

the research questions and objectives. By inputting the research questions and objectives into ChatGPT, researchers can generate an outline that includes the main sections of the research paper.

2. Organizing ideas: ChatGPT can be used to organize ideas and concepts into categories or themes. By inputting the ideas into ChatGPT, researchers can generate a structure that groups related ideas together and helps to organize the research paper.

3. Identifying subtopics: ChatGPT can be used to identify subtopics and subthemes within the research paper. By inputting the main topics into ChatGPT, researchers can generate subtopics and subthemes that help to further organize the research paper.

4. Creating section headings: ChatGPT can be used to create section headings for each section of the research paper. By inputting the main topics and subtopics into ChatGPT, researchers can generate section headings that clearly indicate the content of each section.

5. Providing coherence and flow: ChatGPT can be used to ensure coherence and flow between sections of the research paper. By inputting the section headings and content into ChatGPT, researchers can generate transitions and connective language that help to guide the reader through the research paper.

It is important to note that while ChatGPT can assist with structuring research papers, it should not replace human judgment and expertise. Researchers should use ChatGPT in combination with other writing methods and approaches, and should be aware of the potential biases and limitations of ChatGPT-generated insights. Additionally, researchers should ensure that their use of ChatGPT is ethical and responsible, and that they comply with relevant data protection and privacy laws and regulations.

Writing Abstracts, Introductions, and Conclusions

ChatGPT can be used to assist with writing abstracts, introductions, and conclusions for research papers. Here are some ways that ChatGPT can be leveraged to write these sections:

1. Abstracts: ChatGPT can be used to write abstracts for research papers by summarizing the main findings and conclusions of the paper. By inputting the research paper into ChatGPT, researchers can generate a summary of the paper that includes the purpose, methods, results, and conclusions.

2. Introductions: ChatGPT can be used to write introductions for research papers by providing background information and context for the research question. By inputting the research question into ChatGPT, researchers can generate an introduction that provides a brief overview of the research area and the purpose of the research paper.

3. Conclusions: ChatGPT can be used to write conclusions for research papers by summarizing the main findings and implications of the research. By inputting the research paper into ChatGPT, researchers can generate a conclusion that summarizes the main findings, discusses the implications of the research, and suggests areas for future research.

It is important to note that while ChatGPT can assist with writing abstracts, introductions, and conclusions for research papers, it should not replace human judgment and expertise. Researchers should use ChatGPT in combination with other writing methods and approaches, and should be aware of the potential biases and limitations of ChatGPT-generated insights. Additionally, researchers should ensure that their use of ChatGPT is ethical and responsible, and that they comply with relevant data protection and privacy laws and regulations.

Chapter 8:
Improving Collaboration and Communication

ChatGPT can be used to improve collaboration and communication in research by facilitating idea generation, brainstorming, and knowledge sharing. Here are some ways that ChatGPT can be leveraged to improve collaboration and communication in research:

1. Brainstorming ideas: ChatGPT can be used to facilitate idea generation and brainstorming sessions. Researchers can input keywords or phrases related to the research topic, and ChatGPT can generate a list of potential ideas or directions for the research.
2. Sharing knowledge: ChatGPT can be used to share knowledge and information between researchers. By inputting a question or topic into ChatGPT, researchers can generate summaries of relevant research articles or sources, which can be shared with other team members.
3. Generating feedback: ChatGPT can be used to generate feedback on research ideas or papers. Researchers can input a research idea or paper into ChatGPT, and ChatGPT can provide suggestions for improvements or areas to explore further.
4. Collaborating on writing: ChatGPT can be used to collaborate on writing research papers or reports. Researchers can input a section of the paper into ChatGPT, and ChatGPT can provide suggestions for improving the language, coherence, and flow of the section.
5. Facilitating meetings: ChatGPT can be used to facilitate meetings and discussions between researchers. By inputting discussion topics or questions into ChatGPT, researchers can generate a structure for the meeting and ensure that all relevant topics are covered.

It is important to note that while ChatGPT can improve collaboration and communication in research, it should not replace human interaction and collaboration. Researchers should use ChatGPT in combination with other collaboration tools and methods, and should be aware of the potential biases and limitations of ChatGPT-generated insights. Additionally, researchers should ensure that their use of ChatGPT is ethical and responsible, and that they comply with relevant data protection and privacy laws and regulations.

Collaboration in Research Teams

Collaboration is essential for successful research, and ChatGPT can be used to improve collaboration within research teams. Here are some ways that ChatGPT can be leveraged to facilitate collaboration within research teams:

1. Idea generation: ChatGPT can be used to generate ideas for research projects or papers. Researchers can input keywords or phrases related to the research topic, and ChatGPT can generate a list of potential ideas or directions for the research.
2. Brainstorming sessions: ChatGPT can be used to facilitate idea generation and brainstorming sessions. Researchers can input discussion topics or questions into ChatGPT, and ChatGPT can provide suggestions for topics to explore or directions to take.
3. Sharing knowledge: ChatGPT can be used to share knowledge and information between researchers. By inputting a question or topic into ChatGPT, researchers can generate summaries of relevant research articles or sources, which can be shared with other team members.
4. Collaborating on writing: ChatGPT can be used to collaborate on writing research papers or reports. Researchers can input a section of the paper into ChatGPT, and ChatGPT can provide suggestions for improving the language, coherence, and flow of the section.
5. Providing feedback: ChatGPT can be used to provide feedback on research ideas or papers. Researchers can input a research idea or paper into ChatGPT, and ChatGPT can provide suggestions for improvements or areas to explore further.
6. Facilitating meetings: ChatGPT can be used to facilitate meetings and discussions between researchers. By inputting discussion topics or questions into ChatGPT, researchers can generate a structure for the meeting and ensure that all relevant topics are covered.
7. Collaborating on data analysis: ChatGPT can be used to collaborate on data analysis. Researchers can input data sets into ChatGPT, and ChatGPT can provide suggestions for statistical analyses or visualizations.

It is important to note that while ChatGPT can improve collaboration within research teams, it should not replace human interaction and collaboration. Researchers should use ChatGPT in combination with other collaboration tools and methods, and should be aware of the potential biases and limitations of ChatGPT-generated insights. Additionally, researchers should ensure that their use of ChatGPT is ethical and responsible, and that they comply with relevant data protection and privacy laws and regulations.

Communicating Research Findings to a Non-Expert Audience

Communicating research findings to a non-expert audience can be challenging, but ChatGPT can be used to simplify complex concepts and ideas into plain language. Here are some ways that ChatGPT can be leveraged to communicate research findings to a non-expert audience:

1. Generating summaries: ChatGPT can be used to generate summaries of research findings in plain language. By inputting the research paper or article into ChatGPT, researchers can

generate a summary that explains the key findings and their implications in simple terms.
2. Simplifying concepts: ChatGPT can be used to simplify complex concepts and ideas into plain language. Researchers can input technical terms or jargon into ChatGPT, and ChatGPT can provide simpler explanations or synonyms that are easier for non-experts to understand.
3. Providing examples: ChatGPT can be used to provide examples that illustrate research findings or concepts. By inputting a research finding into ChatGPT, researchers can generate examples that help to make the finding more concrete and accessible to non-experts.
4. Creating visuals: ChatGPT can be used to create visuals that help to communicate research findings. Researchers can input data or statistics into ChatGPT, and ChatGPT can generate visualizations or graphs that illustrate the findings in a clear and accessible way.
5. Identifying practical applications: ChatGPT can be used to identify practical applications of research findings. By inputting research findings into ChatGPT, researchers can generate suggestions for how the findings could be applied in real-world situations, which can help to make the research more relevant and interesting to non-experts.

It is important to note that while ChatGPT can assist with communicating research findings to a non-expert audience, it should not replace human judgment and expertise. Researchers should use ChatGPT in combination with other communication methods and approaches, and should be aware of the potential biases and limitations of ChatGPT-generated insights. Additionally, researchers should ensure that their use of ChatGPT is ethical and responsible, and that they comply with relevant data protection and privacy laws and regulations.

Chapter 9:
Ethical Considerations in AI-Assisted Research

There are several ethical considerations to keep in mind when using AI-assisted research tools, such as ChatGPT. Here are some of the key considerations:

1. Data privacy: Researchers should be aware of data privacy concerns when using AI-assisted research tools. They should ensure that they have obtained informed consent from research participants, and that they are handling and storing data in accordance with relevant data protection and privacy laws and regulations.
2. Bias and fairness: AI algorithms can be biased, and researchers should be aware of the potential for bias in the data sets and models used by AI-assisted research tools. They should take steps to mitigate bias and ensure that the research is fair and equitable.
3. Transparency and accountability: Researchers should be transparent about their use of AI-assisted research tools, and should provide clear explanations of how the tools were used and what their limitations are. They should also be accountable for any decisions made based on AI-generated insights, and should be prepared to explain and justify their decisions to stakeholders.
4. Intellectual property: Researchers should be aware of intellectual property concerns when using AI-assisted research tools. They should ensure that they have the necessary licenses and permissions to use any copyrighted material, and should be aware of the potential for infringement when using AI-generated content.
5. Human oversight: While AI-assisted research tools can be useful, they should not replace human judgment and expertise. Researchers should use AI tools in combination with other research methods and approaches, and should be prepared to exercise human judgment and oversight when interpreting AI-generated insights.

It is important for researchers to be aware of these ethical considerations when using AI-assisted research tools, and to take steps to ensure that their use of these tools is ethical, responsible, and compliant with relevant laws and regulations.

Responsible Use of AI in Research

Responsible use of AI in research involves using AI-assisted research tools in ways that are ethical, transparent, and accountable. Here are some key principles to keep in mind when using AI in research:

1. Transparency: Researchers should be transparent about their use of AI-assisted research tools, including how they were used and what their limitations are. They should be open and honest about the data sets and algorithms used, and should be prepared to explain and justify their decisions to stakeholders.
2. Fairness: Researchers should ensure that their use of AI-assisted research tools is fair and equitable, and should take steps to mitigate bias and ensure that the research is unbiased and non-discriminatory.
3. Human oversight: While AI-assisted research tools can be useful, they should not replace human judgment and expertise. Researchers should use AI tools in combination with other research methods and approaches, and should be prepared to exercise human judgment and oversight when interpreting AI-generated insights.
4. Privacy: Researchers should be aware of data privacy concerns when using AI-assisted research tools. They should ensure that they have obtained informed consent from research participants, and that they are handling and storing data in accordance with relevant data protection and privacy laws and regulations.
5. Accountability: Researchers should be accountable for any decisions made based on AI-generated insights, and should be prepared to explain and justify their decisions to stakeholders. They should also be prepared to address any unintended consequences or negative impacts that arise from their use of AI-assisted research tools.
6. Continuous learning and improvement: Researchers should continually assess and improve their use of AI-assisted research tools, and should be open to feedback and critique from stakeholders. They should be prepared to make changes and adjustments to their research practices in response to new information or insights.

By adhering to these principles, researchers can ensure that their use of AI-assisted research tools is responsible, ethical, and compliant with relevant laws and regulations.

Addressing Bias and Fairness

Addressing bias and ensuring fairness is a critical consideration when using AI in research. Here are some strategies that researchers can use to address bias and ensure fairness:

1. Diverse and representative data: Researchers should ensure that their data sets are diverse and representative, and that they include data from a range of sources and perspectives. This can help to reduce the risk of bias in the data and ensure that the research is fair and equitable.
2. Bias assessment and mitigation: Researchers should assess the potential for bias in their data

sets and models, and should take steps to mitigate any identified biases. This can involve using techniques such as data augmentation, de-biasing algorithms, or sensitivity analysis to ensure that the research is unbiased and fair.

3. Transparency and explainability: Researchers should be transparent about their data sets and algorithms, and should provide clear explanations of how they were used and what their limitations are. This can help to ensure that the research is fair and equitable, and can help to build trust with stakeholders.
4. Human oversight: While AI can be useful in research, it should not replace human judgment and expertise. Researchers should exercise human oversight and judgment when interpreting AI-generated insights, and should be prepared to question and challenge the results to ensure that they are fair and unbiased.
5. Continuous monitoring and evaluation: Researchers should continuously monitor and evaluate their research practices to ensure that they are fair and unbiased. This can involve using metrics such as diversity or representativeness to assess the fairness of the research, and can help to identify areas for improvement or adjustment.

By implementing these strategies, researchers can ensure that their use of AI in research is fair and unbiased, and can help to build trust with stakeholders.

Chapter 10:
Looking Ahead: ChatGPT and the Future of Research

As AI technologies continue to evolve and improve, ChatGPT and other AI-assisted research tools are likely to become more prevalent in the research field. Here are some of the ways that ChatGPT could shape the future of research:

1. Accelerating research: ChatGPT can help to accelerate the research process by automating tasks such as literature reviews, data analysis, and hypothesis generation. This could help researchers to save time and resources, and to focus on higher-level tasks such as interpreting results and generating insights.
2. Enabling new research directions: ChatGPT can help researchers to explore new research directions by suggesting novel hypotheses or research questions that may not have been considered otherwise. This could help to drive innovation and discovery in the research field.
3. Improving research quality: ChatGPT can help to improve the quality of research by identifying potential biases, suggesting alternative approaches, and providing insights that may not have been discovered through traditional research methods.
4. Enhancing collaboration: ChatGPT can facilitate collaboration between researchers by enabling real-time collaboration on research tasks, providing a common language and framework for discussion, and improving the sharing and dissemination of research findings.
5. Broadening accessibility: ChatGPT can help to broaden access to research by simplifying complex concepts and making research findings more accessible to a wider audience. This could help to increase the impact and relevance of research findings and to foster greater public engagement with research.

As with any new technology, there are also potential challenges and risks associated with the use of ChatGPT in research. It will be important for researchers to continue to assess and address these challenges, and to ensure that their use of AI-assisted research tools is responsible, ethical, and compliant with relevant laws and regulations.

Current Trends and Future Predictions

Current trends in the use of AI in research suggest that the technology is becoming increasingly prevalent across a range of disciplines. Here are some of the key trends in the use of AI in research:

1. Natural language processing: Natural language processing (NLP) is a key area of AI research, and is being used to automate tasks such as literature reviews, data extraction, and text summarization. NLP is also being used to develop conversational agents, chatbots, and virtual assistants that can help researchers to find information, answer questions, and complete research tasks more efficiently.
2. Machine learning: Machine learning is another key area of AI research, and is being used to develop predictive models that can help researchers to identify patterns and insights in large data sets. Machine learning is also being used to automate tasks such as image and speech recognition, and to develop recommendation systems that can help researchers to find relevant research articles, journals, and conferences.
3. Data analytics: Data analytics is a growing area of research, and is being used to analyze large and complex data sets from a range of sources. AI tools are being used to help researchers to identify patterns and insights in the data, and to develop predictive models that can help to inform research questions and hypotheses.
4. Collaborative research: Collaborative research is becoming increasingly important, and AI tools are being used to facilitate collaboration between researchers across different disciplines and geographies. AI tools such as ChatGPT can help researchers to communicate more efficiently, share data and research findings, and collaborate on research tasks in real-time.

Looking ahead, it is likely that the use of AI in research will continue to grow, driven by advancements in AI technology, increasing demand for faster and more efficient research methods, and the need for more collaborative and interdisciplinary research. Researchers will need to adapt to these trends by developing new skills and approaches to research, and by ensuring that their use of AI tools is responsible, ethical, and compliant with relevant laws and regulations.

Preparing for an AI-Driven Research Landscape

Preparing for an AI-driven research landscape involves developing the skills, tools, and resources needed to effectively integrate AI into research workflows. Here are some steps that researchers can take to prepare for an AI-driven research landscape:

1. Develop AI skills: Researchers will need to develop skills in areas such as natural language processing, machine learning, and data analytics to effectively use AI tools in their research. This can involve taking courses, attending workshops, and collaborating with experts in these fields.
2. Collaborate across disciplines: AI-driven research is likely to involve collaboration across different disciplines, as researchers with different skill sets work together to develop new AI-

powered research tools and approaches. Researchers should seek out opportunities to collaborate with experts from other fields and to learn from their experiences.

3. Embrace open science: Open science involves making research data, methods, and findings openly available to the wider research community. By embracing open science, researchers can help to foster greater collaboration and innovation in the AI research field, and can contribute to the development of new AI-powered research tools and approaches.

4. Ensure data privacy and security: AI-driven research will involve the collection and use of large amounts of data, and researchers will need to ensure that this data is handled and stored securely and in accordance with relevant data protection and privacy laws and regulations.

5. Be aware of ethical considerations: AI-assisted research tools raise a number of ethical considerations, including issues around bias, privacy, and accountability. Researchers will need to be aware of these issues and take steps to ensure that their use of AI tools is ethical and responsible.

By taking these steps, researchers can prepare themselves for an AI-driven research landscape, and can help to ensure that their use of AI tools is effective, efficient, and responsible.

BOOK 4: "ChatGPT for Educators: Tutoring, Feedback and Personalized Instruction": Discusses the use of ChatGPT in the field of education.

Chapter 1:
Introduction to ChatGPT for Educators

As an educator, you can use ChatGPT to enhance your teaching and improve the learning experience for your students. This guide will provide an introduction to ChatGPT and some of the ways that you can use it in your teaching.

ChatGPT is a type of AI language model that is trained on vast amounts of text data. It uses this data to generate human-like text in response to prompts and questions. ChatGPT is powered by OpenAI, one of the leading research institutions in the field of AI. It is designed to be flexible and adaptable, and can be used in a wide range of applications, including education.

As an educator, you can use ChatGPT to create more engaging and interactive learning experiences for your students. For example, you can use ChatGPT to:

1. Generate discussion prompts: ChatGPT can generate open-ended discussion prompts that can help to stimulate student engagement and critical thinking. These prompts can be used in a variety of contexts, including online forums, class discussions, and group projects.
2. Provide feedback and support: ChatGPT can be used to provide feedback and support to students on their assignments and projects. For example, it can provide automated feedback on grammar and syntax, suggest revisions to written work, and provide guidance on research methods and data analysis.
3. Create personalized learning experiences: ChatGPT can be used to create personalized learning experiences for students by generating tailored recommendations and resources based on their individual learning needs and preferences.
4. Enhance online learning: ChatGPT can be used to enhance online learning experiences by providing interactive and engaging content, such as quizzes, simulations, and games.

By using ChatGPT in your teaching, you can create more engaging and interactive learning experiences for your students, and help to foster critical thinking, creativity, and innovation in the classroom.

Understanding ChatGPT

ChatGPT is a type of AI language model that is trained on vast amounts of text data to generate human-like text in response to prompts and questions. It uses a neural network to process natural language and create coherent, contextually relevant responses.

ChatGPT is based on the GPT (Generative Pre-trained Transformer) architecture, which is a state-of-the-art deep learning model for natural language processing (NLP). The model is pre-trained on large datasets of text, which allows it to learn patterns and relationships in language and make predictions about what words or phrases are likely to follow.

When a user inputs a prompt or question, ChatGPT uses the pre-trained model to generate a response. The model takes into account the context of the prompt and uses its knowledge of language patterns and relationships to generate a response that is relevant and coherent.

ChatGPT is particularly useful for generating creative and innovative ideas, as it can generate responses that are outside of the typical patterns of human language. It can also be used to generate personalized recommendations based on user input, and can be integrated into a variety of applications, including chatbots, virtual assistants, and content creation tools.

Overall, ChatGPT is a powerful tool for natural language processing and can be used in a variety of applications, including education, research, and business.

The Role of AI in Education

Artificial intelligence (AI) is playing an increasingly important role in education, with the potential to transform the way that we teach and learn. Here are some of the key roles that AI is playing in education:

1. Personalized learning: AI can be used to personalize learning experiences for students by creating tailored recommendations and resources based on their individual learning needs and preferences. This can help to improve student engagement and outcomes by providing a more personalized and relevant learning experience.
2. Automated grading and feedback: AI can be used to automate grading and provide feedback on student assignments and projects. This can save teachers time and enable them to provide more detailed and comprehensive feedback to students.
3. Intelligent tutoring systems: AI-powered intelligent tutoring systems can provide personalized guidance and support to students, helping them to master complex concepts and skills. These systems can be tailored to individual students' needs and can adapt to their progress over time.
4. Language learning: AI can be used to improve language learning outcomes by providing immersive and interactive language learning experiences. For example, AI-powered chatbots and virtual assistants can provide conversational language practice and feedback.
5. Research and data analysis: AI can be used to support research and data analysis in education,

helping researchers to identify patterns and insights in large datasets and to develop predictive models that can inform educational policy and practice.

6. Accessibility: AI can be used to improve accessibility in education, providing support for students with disabilities or learning difficulties. For example, AI-powered text-to-speech software can help students with visual impairments to access educational materials.

As AI continues to develop and mature, it is likely that it will play an even greater role in education, helping to create more personalized, engaging, and effective learning experiences for students of all ages and backgrounds. However, it is important to ensure that the use of AI in education is responsible, ethical, and in accordance with relevant laws and regulations.

Chapter 2:
ChatGPT as a Tutoring Tool

ChatGPT can be used as a tutoring tool to provide personalized guidance and support to students. Here are some of the ways that ChatGPT can be used as a tutoring tool:

1. Automated feedback and grading: ChatGPT can be used to automate feedback and grading on student assignments and projects. This can save teachers time and enable them to provide more detailed and comprehensive feedback to students.
2. Personalized learning: ChatGPT can be used to create personalized learning experiences for students by generating tailored recommendations and resources based on their individual learning needs and preferences.
3. Language learning: ChatGPT can be used to improve language learning outcomes by providing immersive and interactive language learning experiences. For example, ChatGPT can generate conversational language practice and feedback.
4. Concept mastery: ChatGPT can help students master complex concepts by generating explanations and examples that are tailored to their individual learning needs.
5. Study skills: ChatGPT can provide students with tips and strategies for effective studying, such as time management techniques and note-taking strategies.
6. Test preparation: ChatGPT can generate practice questions and quizzes to help students prepare for exams and assess their knowledge and understanding of key concepts.

Overall, ChatGPT can be a valuable tool for tutoring, providing personalized support and guidance to students and helping them to achieve their academic goals. However, it is important to ensure that the use of ChatGPT in tutoring is responsible, ethical, and in accordance with relevant laws and regulations.

Reinforcing Classroom Learning with ChatGPT

ChatGPT can be used to reinforce classroom learning by providing additional resources and support to students outside of the classroom. Here are some ways that ChatGPT can be used to reinforce classroom learning:

1. Answering student questions: ChatGPT can be used to answer student questions and provide additional explanations of concepts covered in class. This can help students to reinforce their understanding of key concepts and address any areas of confusion.

2. Generating practice exercises: ChatGPT can generate practice exercises and quizzes for students to complete outside of class. This can help students to reinforce their knowledge and understanding of key concepts and prepare for exams and assessments.
3. Providing additional resources: ChatGPT can provide additional resources such as articles, videos, and interactive simulations that reinforce concepts covered in class. This can help students to deepen their understanding of key concepts and develop a more comprehensive understanding of the subject matter.
4. Personalized learning: ChatGPT can be used to provide personalized learning experiences for students by generating tailored recommendations and resources based on their individual learning needs and preferences. This can help students to reinforce their understanding of key concepts and develop skills and knowledge in areas where they need additional support.
5. Collaboration and discussion: ChatGPT can be used to facilitate collaboration and discussion among students outside of the classroom. For example, it can generate discussion prompts and facilitate online discussions that enable students to share their ideas and perspectives with their peers.

By using ChatGPT to reinforce classroom learning, educators can help students to deepen their understanding of key concepts and develop the skills and knowledge they need to succeed. However, it is important to ensure that the use of ChatGPT is responsible, ethical, and in accordance with relevant laws and regulations.

Supplemental Learning Outside of the Classroom

ChatGPT can be used to provide supplemental learning experiences for students outside of the classroom. Here are some ways that ChatGPT can be used to support learning outside of the classroom:

1. Personalized learning: ChatGPT can provide personalized learning experiences for students by generating tailored recommendations and resources based on their individual learning needs and preferences.
2. Exploring new topics: ChatGPT can generate content on new and emerging topics that are not covered in the classroom. This can help students to explore new areas of interest and develop new skills and knowledge.
3. Interactive learning experiences: ChatGPT can provide interactive learning experiences such as simulations and games that enable students to engage with the material in a more hands-on way.
4. Self-paced learning: ChatGPT can provide self-paced learning experiences that enable students

to learn at their own pace and on their own schedule. This can be particularly beneficial for students who need additional support or who are looking to accelerate their learning.
5. Supplemental resources: ChatGPT can provide additional resources such as articles, videos, and podcasts that provide additional context and support for the material covered in class.

By using ChatGPT to supplement learning outside of the classroom, educators can provide students with additional support and resources to help them achieve their learning goals. However, it is important to ensure that the use of ChatGPT is responsible, ethical, and in accordance with relevant laws and regulations. Additionally, educators should ensure that students have access to the necessary technology and resources to effectively engage with ChatGPT and other AI-powered learning tools.

Chapter 3:
Personalized Instruction with ChatGPT

ChatGPT can provide personalized instruction by generating tailored recommendations and resources based on each student's individual learning needs and preferences. Here are some ways that ChatGPT can be used to provide personalized instruction:

1. Assessing student knowledge and understanding: ChatGPT can be used to assess each student's knowledge and understanding of key concepts, helping educators to identify areas where each student needs additional support.
2. Generating tailored resources: ChatGPT can generate tailored resources such as articles, videos, and interactive simulations that are tailored to each student's individual learning needs and preferences. This can help students to reinforce their understanding of key concepts and develop skills and knowledge in areas where they need additional support.
3. Adaptive learning: ChatGPT can be used to provide adaptive learning experiences that adjust to each student's progress and performance. This can help students to develop skills and knowledge at their own pace and on their own schedule.
4. Personalized feedback and guidance: ChatGPT can provide personalized feedback and guidance on student assignments and projects, helping students to improve their work and develop their skills and knowledge.
5. Tracking progress and performance: ChatGPT can track each student's progress and performance, providing educators with insights into each student's learning journey and enabling them to adjust their instruction accordingly.

By providing personalized instruction with ChatGPT, educators can help each student to achieve their full potential by tailoring their instruction to their individual learning needs and preferences. However, it is important to ensure that the use of ChatGPT is responsible, ethical, and in accordance with relevant laws and regulations. Additionally, educators should ensure that students have access to the necessary technology and resources to effectively engage with ChatGPT and other AI-powered learning tools.

Tailoring Instruction to Individual Learning Styles

ChatGPT can be used to tailor instruction to individual learning styles by generating resources and recommendations that are tailored to each student's preferred learning style. Here are some ways that ChatGPT can be used to tailor instruction to individual learning styles:

1. Visual learners: ChatGPT can generate visual resources such as videos, infographics, and diagrams that are tailored to visual learners. This can help visual learners to reinforce their understanding of key concepts and develop their skills and knowledge.
2. Auditory learners: ChatGPT can generate audio resources such as podcasts and recordings of lectures that are tailored to auditory learners. This can help auditory learners to reinforce their understanding of key concepts and develop their skills and knowledge.
3. Kinesthetic learners: ChatGPT can generate interactive resources such as simulations and games that are tailored to kinesthetic learners. This can help kinesthetic learners to engage with the material in a more hands-on way and develop their skills and knowledge.
4. Reading and writing learners: ChatGPT can generate text-based resources such as articles, textbooks, and writing assignments that are tailored to reading and writing learners. This can help these learners to reinforce their understanding of key concepts and develop their reading and writing skills.
5. Adaptive learning: ChatGPT can provide adaptive learning experiences that adjust to each student's preferred learning style. This can help students to develop skills and knowledge in a way that is comfortable and effective for them.

By tailoring instruction to individual learning styles with ChatGPT, educators can help each student to learn in a way that is comfortable and effective for them, ultimately leading to better learning outcomes. However, it is important to ensure that the use of ChatGPT is responsible, ethical, and in accordance with relevant laws and regulations. Additionally, educators should ensure that students have access to the necessary technology and resources to effectively engage with ChatGPT and other AI-powered learning tools.

Adapting to Students' Pace and Level of Understanding

ChatGPT can be used to adapt instruction to each student's pace and level of understanding by providing personalized learning experiences that adjust to each student's progress and performance. Here are some ways that ChatGPT can be used to adapt instruction:

1. Assessing student knowledge and understanding: ChatGPT can be used to assess each student's knowledge and understanding of key concepts, helping educators to identify areas where each student needs additional support.
2. Providing personalized resources: ChatGPT can generate personalized resources such as articles, videos, and interactive simulations that are tailored to each student's individual learning needs and preferences. This can help students to reinforce their understanding of key concepts and

develop skills and knowledge in areas where they need additional support.

3. Adaptive learning: ChatGPT can provide adaptive learning experiences that adjust to each student's pace and level of understanding. This can help students to develop skills and knowledge at their own pace and on their own schedule.

4. Personalized feedback and guidance: ChatGPT can provide personalized feedback and guidance on student assignments and projects, helping students to improve their work and develop their skills and knowledge.

5. Tracking progress and performance: ChatGPT can track each student's progress and performance, providing educators with insights into each student's learning journey and enabling them to adjust their instruction accordingly.

By adapting instruction with ChatGPT, educators can help each student to learn in a way that is comfortable and effective for them, ultimately leading to better learning outcomes. However, it is important to ensure that the use of ChatGPT is responsible, ethical, and in accordance with relevant laws and regulations. Additionally, educators should ensure that students have access to the necessary technology and resources to effectively engage with ChatGPT and other AI-powered learning tools.

Chapter 4:
Feedback and Assessment

ChatGPT can be used to provide feedback and assessment on student work, helping educators to provide personalized guidance and support to each student. Here are some ways that ChatGPT can be used for feedback and assessment:

1. Automated grading: ChatGPT can be used to automate the grading of multiple-choice and short-answer questions, saving educators time and enabling them to provide faster feedback to students.
2. Personalized feedback: ChatGPT can provide personalized feedback on student assignments and projects, helping students to understand their strengths and weaknesses and improve their work.
3. Peer assessment: ChatGPT can facilitate peer assessment, allowing students to provide feedback on each other's work and learn from each other's perspectives.
4. Performance tracking: ChatGPT can track each student's progress and performance, providing educators with insights into each student's learning journey and enabling them to adjust their instruction and feedback accordingly.
5. Rubric-based assessment: ChatGPT can use rubrics to provide structured and consistent feedback on student work, enabling educators to provide feedback that is objective and aligned with learning objectives.

By using ChatGPT for feedback and assessment, educators can provide personalized guidance and support to each student, ultimately leading to better learning outcomes. However, it is important to ensure that the use of ChatGPT is responsible, ethical, and in accordance with relevant laws and regulations. Additionally, educators should ensure that students have access to the necessary technology and resources to effectively engage with ChatGPT and other AI-powered learning tools.

Using ChatGPT for Instant Feedback

ChatGPT can be used to provide instant feedback to students, enabling them to receive feedback on their work in real-time. Here are some ways that ChatGPT can be used for instant feedback:

1. Multiple-choice quizzes: ChatGPT can be used to create multiple-choice quizzes that provide instant feedback to students based on their answers. This can help students to reinforce their

understanding of key concepts and identify areas where they need additional support.

2. Grammar and spelling checks: ChatGPT can be used to check the grammar and spelling of student writing assignments, providing instant feedback on errors and helping students to improve their writing skills.

3. Interactive simulations: ChatGPT can generate interactive simulations that provide instant feedback to students based on their actions. This can help students to learn through experimentation and trial-and-error.

4. Real-time chatbots: ChatGPT can be used to create real-time chatbots that provide instant feedback to students as they ask questions and seek guidance. This can help students to receive personalized support in a timely manner.

5. Voice and speech recognition: ChatGPT can use voice and speech recognition to provide instant feedback to students on their pronunciation and speaking skills. This can help students to improve their language skills and develop their confidence.

By providing instant feedback with ChatGPT, educators can help students to learn more effectively and efficiently by enabling them to receive feedback in real-time. However, it is important to ensure that the use of ChatGPT is responsible, ethical, and in accordance with relevant laws and regulations. Additionally, educators should ensure that students have access to the necessary technology and resources to effectively engage with ChatGPT and other AI-powered learning tools.

Creating and Grading Assessments with AI

ChatGPT can be used to create and grade assessments in a way that is fast, efficient, and accurate. Here are some ways that ChatGPT can be used for creating and grading assessments with AI:

1. Automated question generation: ChatGPT can be used to automatically generate questions for assessments, including multiple-choice, true/false, and short-answer questions. This can save educators time and enable them to create assessments more efficiently.

2. Automatic grading: ChatGPT can be used to automatically grade assessments, including multiple-choice and short-answer questions. This can save educators time and enable them to provide faster feedback to students.

3. Essay grading: ChatGPT can be used to grade essays using natural language processing techniques. This can enable educators to provide more accurate and consistent feedback to students.

4. Adaptive testing: ChatGPT can be used to create adaptive tests that adjust the difficulty of questions based on each student's performance. This can provide a more accurate measure of

each student's knowledge and abilities.
5. Item analysis: ChatGPT can be used to analyze each item in an assessment, providing educators with insights into the difficulty level, discrimination, and effectiveness of each question.

By using ChatGPT to create and grade assessments, educators can save time, improve accuracy, and provide more consistent feedback to students. However, it is important to ensure that the use of ChatGPT is responsible, ethical, and in accordance with relevant laws and regulations. Additionally, educators should ensure that students have access to the necessary technology and resources to effectively engage with ChatGPT and other AI-powered learning tools.

Chapter 5:
ChatGPT in Lesson Planning

ChatGPT can be used to assist educators in lesson planning by providing ideas, resources, and feedback. Here are some ways that ChatGPT can be used for lesson planning:

1. Generating lesson plan ideas: ChatGPT can be used to generate ideas for lesson plans, providing educators with inspiration and suggestions for new and engaging activities and exercises.
2. Providing learning resources: ChatGPT can generate resources such as articles, videos, and interactive simulations that are relevant to each lesson plan. This can help educators to provide students with a more engaging and interactive learning experience.
3. Personalizing lesson plans: ChatGPT can be used to personalize lesson plans based on each student's individual learning needs and preferences. This can help educators to provide more targeted and effective instruction.
4. Assessment and feedback: ChatGPT can provide feedback on each lesson plan, helping educators to improve their instructional design and better meet the needs of their students.
5. Collaboration and communication: ChatGPT can facilitate collaboration and communication between educators, enabling them to share ideas, resources, and feedback with each other.

By using ChatGPT for lesson planning, educators can save time, improve the quality of their instruction, and provide a more engaging and effective learning experience for their students. However, it is important to ensure that the use of ChatGPT is responsible, ethical, and in accordance with relevant laws and regulations. Additionally, educators should ensure that students have access to the necessary technology and resources to effectively engage with ChatGPT and other AI-powered learning tools.

Curriculum Design with ChatGPT

ChatGPT can be used to assist in curriculum design by providing ideas, resources, and feedback. Here are some ways that ChatGPT can be used for curriculum design:

1. Generating curriculum ideas: ChatGPT can be used to generate ideas for curriculum, providing educators with inspiration and suggestions for new and engaging topics, units, and activities.
2. Providing learning resources: ChatGPT can generate resources such as articles, videos, and interactive simulations that are relevant to each curriculum, helping educators to provide students with a more engaging and interactive learning experience.

3. Personalizing curriculum: ChatGPT can be used to personalize curriculum based on each student's individual learning needs and preferences, helping educators to provide more targeted and effective instruction.
4. Assessment and feedback: ChatGPT can provide feedback on each curriculum, helping educators to improve their instructional design and better meet the needs of their students.
5. Collaboration and communication: ChatGPT can facilitate collaboration and communication between educators, enabling them to share ideas, resources, and feedback with each other.

By using ChatGPT for curriculum design, educators can save time, improve the quality of their instruction, and provide a more engaging and effective learning experience for their students. However, it is important to ensure that the use of ChatGPT is responsible, ethical, and in accordance with relevant laws and regulations. Additionally, educators should ensure that students have access to the necessary technology and resources to effectively engage with ChatGPT and other AI-powered learning tools.

Innovative Lesson Ideas and Resources

ChatGPT can be used to generate innovative lesson ideas and resources for educators. Here are some ways that ChatGPT can be used for innovative lesson ideas and resources:

1. Interactive simulations: ChatGPT can generate interactive simulations that allow students to engage with complex concepts in a visual and interactive way.
2. Augmented and virtual reality: ChatGPT can provide ideas and resources for using augmented and virtual reality in the classroom, enabling students to explore and interact with virtual environments.
3. Gamification: ChatGPT can provide ideas and resources for incorporating game elements into the classroom, making learning more engaging and interactive.
4. Personalized learning: ChatGPT can provide ideas and resources for personalized learning, helping educators to adapt instruction to meet the unique needs and preferences of each student.
5. Project-based learning: ChatGPT can provide ideas and resources for project-based learning, enabling students to work on real-world problems and develop critical thinking skills.

By using ChatGPT to generate innovative lesson ideas and resources, educators can provide students with a more engaging and effective learning experience. However, it is important to ensure that the use of ChatGPT is responsible, ethical, and in accordance with relevant laws and regulations. Additionally, educators should ensure that students have access to the necessary technology and resources to effectively engage with ChatGPT and other AI-powered learning tools.

Chapter 6:
Encouraging Student Engagement

ChatGPT can be used to encourage student engagement in several ways. Here are some strategies:

1. Personalized learning: ChatGPT can help educators to personalize learning by providing each student with content and activities that are tailored to their individual learning needs and preferences.
2. Interactive simulations: ChatGPT can generate interactive simulations that allow students to engage with complex concepts in a visual and interactive way, making learning more engaging and interactive.
3. Gamification: ChatGPT can help educators to incorporate game elements into the classroom, making learning more fun and engaging for students.
4. Collaborative learning: ChatGPT can facilitate collaborative learning by enabling students to work together on projects and assignments, encouraging them to share ideas and learn from each other.
5. Real-world applications: ChatGPT can help educators to connect classroom learning to real-world applications, helping students to see the relevance and importance of what they are learning.
6. Feedback and assessment: ChatGPT can provide students with immediate feedback on their work, helping them to identify areas for improvement and stay motivated.

By using ChatGPT to encourage student engagement, educators can create a more dynamic and interactive learning environment, helping students to stay motivated and engaged in the learning process. However, it is important to ensure that the use of ChatGPT is responsible, ethical, and in accordance with relevant laws and regulations. Additionally, educators should ensure that students have access to the necessary technology and resources to effectively engage with ChatGPT and other AI-powered learning tools.

Using ChatGPT to Facilitate Discussions

ChatGPT can be used to facilitate discussions in the classroom by generating prompts, questions, and topics for discussion. Here are some strategies:

1. Open-ended questions: ChatGPT can generate open-ended questions that encourage students

to think critically and express their opinions.
2. Relevant topics: ChatGPT can generate relevant and timely topics for discussion, helping students to connect classroom learning to current events and real-world issues.
3. Diverse perspectives: ChatGPT can provide prompts and questions that encourage students to consider diverse perspectives and explore different points of view.
4. Scaffolding: ChatGPT can generate prompts and questions that scaffold discussion, helping students to build on each other's ideas and deepen their understanding of the topic.
5. Follow-up questions: ChatGPT can generate follow-up questions that encourage students to elaborate on their ideas and explore the topic in greater depth.

By using ChatGPT to facilitate discussions in the classroom, educators can create a more dynamic and interactive learning environment, helping students to develop critical thinking and communication skills. However, it is important to ensure that the use of ChatGPT is responsible, ethical, and in accordance with relevant laws and regulations. Additionally, educators should ensure that students have access to the necessary technology and resources to effectively engage with ChatGPT and other AI-powered learning tools.

Enhancing Student Participation and Collaboration

ChatGPT can be used to enhance student participation and collaboration in several ways. Here are some strategies:

1. Collaborative projects: ChatGPT can generate ideas for collaborative projects, enabling students to work together on meaningful and engaging projects.
2. Peer feedback: ChatGPT can provide prompts and questions that encourage peer feedback, helping students to improve their work and learn from each other.
3. Group discussion prompts: ChatGPT can generate discussion prompts that encourage group participation and collaboration, enabling students to share their ideas and perspectives with each other.
4. Personalized learning: ChatGPT can help educators to personalize learning by providing each student with content and activities that are tailored to their individual learning needs and preferences, making learning more engaging and effective.
5. Gamification: ChatGPT can provide ideas and resources for incorporating game elements into the classroom, making learning more fun and engaging for students.

By using ChatGPT to enhance student participation and collaboration, educators can create a more dynamic and interactive learning environment, helping students to develop important collaboration and communication skills. However, it is important to ensure that the use of ChatGPT is responsible, ethical, and in accordance with relevant laws and regulations. Additionally, educators should ensure that students have access to the necessary technology and resources to effectively engage with ChatGPT and other AI-powered learning tools.

Chapter 7:
ChatGPT for Professional Development

ChatGPT can be used for professional development in several ways. Here are some strategies:

1. Continuing education: ChatGPT can generate ideas and resources for continuing education, helping professionals to stay up-to-date with the latest trends and best practices in their field.
2. Skill development: ChatGPT can provide resources and guidance for developing new skills, enabling professionals to expand their knowledge and capabilities.
3. Professional networking: ChatGPT can provide guidance and support for building professional networks, connecting professionals with others in their field and facilitating collaboration and learning.
4. Leadership development: ChatGPT can provide resources and guidance for developing leadership skills, enabling professionals to become more effective leaders and mentors.
5. Career planning: ChatGPT can provide guidance and resources for career planning, helping professionals to set goals and create a path for advancement in their field.

By using ChatGPT for professional development, professionals can stay competitive in their field, expand their knowledge and skills, and advance their careers. However, it is important to ensure that the use of ChatGPT is responsible, ethical, and in accordance with relevant laws and regulations. Additionally, professionals should ensure that they have access to the necessary technology and resources to effectively engage with ChatGPT and other AI-powered professional development tools.

Lifelong Learning for Educators

Lifelong learning is critical for educators, as it enables them to stay up-to-date with the latest trends and best practices in their field, as well as develop new skills and knowledge that can improve their teaching practice. ChatGPT can be a useful tool for educators who are looking to engage in lifelong learning. Here are some strategies:

1. Self-directed learning: ChatGPT can generate resources and ideas for self-directed learning, enabling educators to explore areas of interest and deepen their knowledge in specific areas.
2. Professional development: ChatGPT can provide resources and guidance for professional development, helping educators to stay up-to-date with the latest trends and best practices in their field.

3. Collaboration and networking: ChatGPT can provide resources and guidance for collaboration and networking, enabling educators to connect with others in their field and share ideas and best practices.
4. Reflective practice: ChatGPT can provide prompts and questions that encourage reflective practice, helping educators to evaluate their teaching practice and identify areas for improvement.
5. Skill development: ChatGPT can provide resources and guidance for developing new skills, enabling educators to expand their knowledge and capabilities.

By using ChatGPT for lifelong learning, educators can improve their teaching practice, stay up-to-date with the latest trends and best practices in their field, and continue to grow and develop as professionals. However, it is important to ensure that the use of ChatGPT is responsible, ethical, and in accordance with relevant laws and regulations. Additionally, educators should ensure that they have access to the necessary technology and resources to effectively engage with ChatGPT and other AI-powered learning tools.

Exploring New Teaching Strategies with AI

AI can be a valuable tool for educators looking to explore new teaching strategies. Here are some strategies for using AI in the classroom:

1. Personalized learning: AI can be used to personalize learning by providing each student with content and activities that are tailored to their individual learning needs and preferences, making learning more engaging and effective.
2. Interactive learning: AI can be used to create interactive learning experiences, enabling students to engage with content in new and exciting ways.
3. Gamification: AI can provide ideas and resources for incorporating game elements into the classroom, making learning more fun and engaging for students.
4. Collaborative learning: AI can be used to facilitate collaborative learning, enabling students to work together on meaningful and engaging projects.
5. Innovative assessments: AI can be used to develop innovative assessments, enabling educators to evaluate student learning in new and creative ways.

By using AI to explore new teaching strategies, educators can create a more dynamic and interactive learning environment, helping students to develop important collaboration and communication skills. However, it is important to ensure that the use of AI is responsible, ethical, and in accordance with relevant laws and regulations. Additionally, educators should ensure that students have access to the necessary technology and resources to effectively engage with AI-powered learning tools.

Chapter 8:
Parent-Teacher Communication

Effective communication between parents and teachers is essential for student success. ChatGPT can be a useful tool for facilitating parent-teacher communication. Here are some strategies:

1. Regular updates: ChatGPT can be used to provide regular updates on student progress and classroom activities, enabling parents to stay informed and engaged in their child's education.
2. Two-way communication: ChatGPT can be used to facilitate two-way communication between parents and teachers, enabling both parties to share information and ask questions.
3. Parent-teacher conferences: ChatGPT can be used to schedule and coordinate parent-teacher conferences, making it easier for parents and teachers to find a mutually convenient time to meet.
4. Resource sharing: ChatGPT can be used to share resources with parents, such as educational materials and resources for supporting their child's learning at home.
5. Translation: ChatGPT can be used to translate messages between parents and teachers who speak different languages, enabling effective communication regardless of language barriers.

By using ChatGPT to facilitate parent-teacher communication, educators can improve parental involvement in their child's education, leading to better student outcomes. However, it is important to ensure that the use of ChatGPT is responsible, ethical, and in accordance with relevant laws and regulations. Additionally, educators should ensure that parents have access to the necessary technology and resources to effectively engage with ChatGPT and other AI-powered communication tools.

Using ChatGPT to Keep Parents Informed

ChatGPT can be used by educators to keep parents informed and engaged in their child's education. Here are some strategies:

1. Regular updates: ChatGPT can be used to provide regular updates on student progress and classroom activities, enabling parents to stay informed about their child's academic performance and engagement.
2. Student portfolios: ChatGPT can be used to create digital student portfolios that showcase student work, progress, and achievements. These portfolios can be shared with parents to keep

them informed about their child's development.
3. Homework and assignments: ChatGPT can be used to share homework assignments and project guidelines with parents, enabling them to support their child's learning at home.
4. Event notifications: ChatGPT can be used to notify parents about upcoming school events, such as parent-teacher conferences, school assemblies, and extracurricular activities.
5. Access to learning resources: ChatGPT can be used to share learning resources with parents, such as educational articles, videos, and podcasts that support their child's learning at home.

By using ChatGPT to keep parents informed, educators can build stronger partnerships with families and create a more supportive learning environment for students. However, it is important to ensure that the use of ChatGPT is responsible, ethical, and in accordance with relevant laws and regulations. Additionally, educators should ensure that parents have access to the necessary technology and resources to effectively engage with ChatGPT and other AI-powered communication tools.

Automating Routine Communications

ChatGPT can be used by educators to automate routine communications, such as reminders and announcements, enabling them to save time and focus on other important tasks. Here are some strategies:

1. Attendance notifications: ChatGPT can be used to send automated attendance notifications to parents, letting them know when their child is absent from school.
2. Assignment reminders: ChatGPT can be used to send automated assignment reminders to students and parents, helping them to stay on track with homework and projects.
3. Event announcements: ChatGPT can be used to send automated event announcements to parents, notifying them about upcoming school events such as field trips and parent-teacher conferences.
4. Grading updates: ChatGPT can be used to send automated grading updates to students and parents, enabling them to stay informed about academic performance.
5. Routine communication: ChatGPT can be used to automate routine communication, such as announcements and reminders, enabling educators to save time and focus on other important tasks.

By using ChatGPT to automate routine communications, educators can streamline their workflow and reduce the administrative burden of communicating with students and parents. However, it is important to ensure that the use of ChatGPT is responsible, ethical, and in accordance with relevant laws and regulations. Additionally, educators should ensure that students and parents have access to the necessary technology and resources to effectively engage with ChatGPT and other AI-powered communication tools.

Chapter 9:
Ethical Considerations in AI-Assisted Education

There are several ethical considerations that educators should be aware of when using AI-assisted education tools, such as ChatGPT. Here are some of the key considerations:

1. Privacy: It is important to protect the privacy and security of student data when using AI-assisted education tools. Educators should ensure that student data is kept secure and that they have obtained the necessary consent from parents and guardians.
2. Bias and fairness: AI-assisted education tools may have inherent biases that can lead to unfair outcomes. Educators should be aware of the potential for bias and take steps to mitigate it, such as ensuring that the data used to train the AI is diverse and representative.
3. Transparency: Educators should be transparent about their use of AI-assisted education tools and explain to students and parents how they work and what data is being collected.
4. Accountability: Educators should take responsibility for the decisions made by AI-assisted education tools and be accountable for any errors or biases that may arise.
5. Human oversight: While AI can be a valuable tool in education, it should never replace the role of human educators. Educators should ensure that AI is used as a supplement to, rather than a replacement for, human instruction and guidance.

By being aware of these ethical considerations and taking steps to address them, educators can ensure that they are using AI-assisted education tools in a responsible and ethical manner. Additionally, it is important for educators to stay up-to-date on developments in AI and education, as well as relevant laws and regulations, to ensure that their use of AI is in accordance with best practices and ethical standards.

Responsible Use of AI in the Classroom

To ensure responsible use of AI in the classroom, educators should follow these guidelines:

1. Ensure Transparency: Educators should ensure that students are informed about the use of AI in the classroom, including how it works and how it is being used.
2. Protect Student Privacy: Educators should take measures to ensure that student data is protected and that they have obtained the necessary consent from parents and guardians.
3. Monitor for Bias: Educators should be aware of the potential for bias in AI systems and take

steps to mitigate it, such as using diverse and representative datasets.
4. Encourage Critical Thinking: Educators should encourage students to think critically about the use of AI and to question its limitations and potential biases.
5. Use AI as a Tool: AI should be used as a tool to support teaching and learning, rather than as a replacement for human instruction.
6. Continuously Monitor and Evaluate: Educators should continuously monitor and evaluate the use of AI in the classroom to ensure that it is achieving its intended goals and that any issues are addressed.

By following these guidelines, educators can ensure that they are using AI in the classroom in a responsible and ethical manner. Additionally, it is important for educators to stay up-to-date on developments in AI and education, as well as relevant laws and regulations, to ensure that their use of AI is in accordance with best practices and ethical standards.

Addressing Equity and Access Issues

Addressing equity and access issues is an important consideration when using AI in the classroom. Here are some strategies that educators can use to ensure that all students have equal access to AI-assisted education:

1. Provide access to technology: To ensure that all students have equal access to AI-assisted education, educators should provide access to the necessary technology, such as computers and internet access.
2. Use inclusive data sets: AI algorithms can be biased if the data sets they are trained on are not diverse and inclusive. Educators should ensure that the data sets used to train AI are diverse and representative of all students.
3. Design for accessibility: When using AI-assisted education tools, educators should ensure that they are designed for accessibility, such as providing options for closed captioning or text-to-speech.
4. Be mindful of cultural biases: Educators should be mindful of cultural biases that may be inherent in AI systems and take steps to address them, such as using culturally responsive teaching practices.
5. Encourage student feedback: Educators should encourage students to provide feedback on the use of AI-assisted education tools, particularly if they feel that they are not accessible or inclusive.

By using these strategies, educators can help ensure that all students have equal access to AI-assisted education tools and that AI is used in an equitable and inclusive manner. Additionally, educators should continuously monitor and evaluate the use of AI in the classroom to ensure that it is achieving its intended goals and that any issues are addressed.

Chapter 10:
Looking Ahead: ChatGPT and the Future of Education

The use of ChatGPT and other AI technologies in education is still in its early stages, but it has the potential to transform the way we teach and learn. Here are some potential future developments for ChatGPT and AI in education:

1. Personalized Learning: AI can be used to personalize learning experiences for individual students, taking into account their learning styles, interests, and abilities. ChatGPT could be used to generate personalized learning plans, activities, and assessments.
2. Adaptive Learning: AI can be used to adapt learning materials and activities in real-time based on a student's progress and understanding of the material. ChatGPT could be used to generate questions and feedback based on a student's performance.
3. Virtual Assistants: ChatGPT could be used as virtual assistants for educators, answering students' questions and providing real-time feedback on their work.
4. Intelligent Tutoring Systems: AI could be used to create intelligent tutoring systems that provide individualized instruction and feedback to students. ChatGPT could be used to generate explanations and examples tailored to a student's needs.
5. Lifelong Learning: AI can be used to provide personalized, lifelong learning opportunities to people of all ages and backgrounds. ChatGPT could be used to generate learning materials and activities for adult learners and those seeking professional development.

As these developments take place, it will be important for educators to consider the ethical implications of AI in education, such as privacy, bias, and fairness. Additionally, it will be important to ensure that AI is used to supplement, rather than replace, human instruction and to ensure that all students have equal access to AI-assisted education. By doing so, we can ensure that ChatGPT and other AI technologies are used in a responsible and equitable manner to improve education for all.

Current Trends and Future Predictions

Here are some current trends and future predictions related to the use of ChatGPT and AI in education:

1. Increasing Use of ChatGPT in Education: ChatGPT and other AI technologies are becoming increasingly prevalent in education, as educators seek to leverage these tools to improve teaching and learning.

2. Integration with Learning Management Systems: AI technologies are being integrated with learning management systems to provide personalized learning experiences and automate routine tasks such as grading and feedback.
3. Expansion of Virtual and Remote Learning: The COVID-19 pandemic has accelerated the adoption of virtual and remote learning, and AI technologies are playing an increasingly important role in facilitating these modes of learning.
4. Advancements in Natural Language Processing: As natural language processing technologies continue to improve, ChatGPT and other AI tools will become more effective at generating natural language responses and engaging in more complex interactions with students.
5. Continued Focus on Equity and Access: There will be an increased focus on ensuring that AI technologies are used in a way that is equitable and inclusive, with efforts to mitigate biases and ensure that all students have equal access to these tools.
6. Growing Importance of Data Privacy: As AI tools collect and analyze more student data, there will be a growing emphasis on data privacy and security, with efforts to protect student information and ensure that it is used in a responsible manner.

Overall, the future of ChatGPT and AI in education is promising, with the potential to transform teaching and learning and improve access to education for all. As with any technology, it will be important to ensure that AI is used in an ethical and responsible manner, with a focus on equity, inclusion, and student privacy.

Preparing for an AI-Driven Educational Landscape

As AI technologies continue to advance and become more prevalent in education, there are several ways in which educators can prepare for an AI-driven educational landscape:

1. Learn about AI technologies: Educators should take the time to learn about the various AI technologies that are available and how they can be used in education. This includes understanding the capabilities and limitations of these tools, as well as the ethical considerations associated with their use.
2. Explore new teaching methods: AI technologies can enable new teaching methods that were not previously possible, such as personalized learning and adaptive instruction. Educators should explore these new teaching methods and think about how they can be integrated into their classroom practices.
3. Develop new skills: As AI technologies become more prevalent, educators will need to develop new skills related to the use of these tools. This may include skills related to data analysis,

programming, and natural language processing.
4. Embrace lifelong learning: Lifelong learning will become increasingly important as AI technologies continue to evolve and new tools are developed. Educators should embrace lifelong learning and seek out opportunities to continue developing their skills and knowledge.
5. Foster a culture of innovation: To fully leverage the potential of AI technologies in education, educators must foster a culture of innovation and experimentation. This includes encouraging experimentation with new teaching methods, as well as providing opportunities for students to explore and learn with AI technologies.

Overall, educators must be willing to adapt to the changing educational landscape and embrace new technologies such as ChatGPT and other AI tools. By doing so, we can ensure that our students are well-prepared for the future and have the skills and knowledge they need to succeed in a rapidly-evolving world.

The best PROMPTs for the STUDENTS and TEACHERS:

Write in GPT Chat this prompt exactly as it is written below. Then try changing the terms you find in the " " to get the work that works best for you. Remember that in case you are a students or teachers, this prompt is PERFECT for you, it is a very powerful tool make good use of it! Prompt:

You are a teacher and will act like one. You will write a lesson with a duration of "one" hour about "environmental pollution" invented by you, with a "formal" theme, for an audience of "students" with age about "10-12" which aims to tell, explain and reflect. You will use an "exciting" writing style, an "engaging" sentiment, and a formal communicative register.

You are a student with age about "18-19" and will act like one. You will write a research about "environmental pollution", with a "formal" theme, for an audience of "teacher" which aims to tell, explain and reflect and take a good vote. You will use an "professional" writing style, an "engaging" sentiment, and a formal communicative register.

BOOK 5: "ChatGPT for Writers: Idea Generation, Correction and Editing": Explores how ChatGPT can be a powerful tool for writers.

Chapter 1:
Introduction to ChatGPT for Writers

Welcome to an introduction to ChatGPT for writers. ChatGPT is an AI language model that has been trained on a massive amount of text data and is capable of generating human-like text. It can be a useful tool for writers in a variety of ways, from generating ideas to improving the quality of writing.

In this guide, we will explore some of the ways that writers can use ChatGPT to enhance their craft. We will discuss how ChatGPT can be used for idea generation, improving writing skills, and automating routine tasks. We will also touch on some of the ethical considerations that come with using AI in writing.

Whether you're a novelist, journalist, blogger, or technical writer, ChatGPT has the potential to help you take your writing to the next level. So let's dive in and explore the possibilities!

Understanding ChatGPT

ChatGPT is an artificial intelligence language model developed by OpenAI. It is based on the GPT (Generative Pre-trained Transformer) architecture and is one of the most powerful and sophisticated language models currently available.

ChatGPT is trained on a massive amount of text data from a variety of sources, including books, articles, and websites. This training allows the model to generate human-like text and respond to a wide range of prompts and questions.

To use ChatGPT, you simply input a prompt or question and the model will generate a response. The response can be customized by adjusting the length and other settings, and can be used for a variety of purposes, from generating ideas to improving the quality of writing.

ChatGPT has been used in a wide range of applications, including language translation, chatbots, and content creation. In the context of writing, ChatGPT can be used to generate ideas for stories or articles, help writers overcome writer's block, and even generate entire pieces of writing.

Overall, ChatGPT is a powerful tool for writers that has the potential to greatly enhance the writing process and produce high-quality content.

The Role of AI in Writing

The role of artificial intelligence (AI) in writing is rapidly evolving and expanding. AI-powered tools like

ChatGPT are becoming increasingly sophisticated, and are now capable of generating human-like text, identifying errors in grammar and syntax, and even providing feedback on the overall quality of writing.

One of the primary roles of AI in writing is to assist writers with routine tasks, such as proofreading and editing. AI-powered tools can quickly identify errors in spelling, grammar, and syntax, and suggest corrections or improvements. This can save writers a significant amount of time and improve the overall quality of their writing.

AI can also be used to generate ideas and overcome writer's block. By inputting a prompt or question, AI language models like ChatGPT can generate a wide range of responses and ideas, providing writers with new perspectives and insights.

Another important role of AI in writing is in content creation. AI-powered tools can generate high-quality content on a variety of topics, from news articles to product descriptions. This has the potential to greatly streamline the content creation process and allow writers to focus on more complex and creative tasks.

Overall, the role of AI in writing is to enhance the writing process, providing writers with new tools and capabilities that can improve the quality and efficiency of their work. While there are still some limitations to the technology, AI-powered writing tools like ChatGPT have the potential to revolutionize the way we write and create content.

Chapter 2:
Idea Generation with ChatGPT

One of the key benefits of using ChatGPT for writers is its ability to generate new ideas and perspectives. Here are some ways that ChatGPT can be used for idea generation:

1. Prompt generation: ChatGPT can generate a variety of prompts and questions based on different topics or themes. These prompts can be used to inspire new ideas and generate fresh perspectives.
2. Story generation: ChatGPT can be used to generate story ideas, plot twists, and character traits. By inputting a prompt such as "a protagonist who has lost their memory," ChatGPT can generate a variety of potential storylines that writers can explore.
3. Topic exploration: ChatGPT can also be used to explore new topics and areas of interest. By inputting a topic or keyword, ChatGPT can generate related ideas and concepts that writers can use to develop new articles or pieces of writing.
4. Content research: ChatGPT can also be used to help writers research and gather information for their writing. By inputting a research question, ChatGPT can generate a variety of sources and ideas that can be used to support the writer's work.

Overall, ChatGPT can be a valuable tool for writers looking to generate new ideas and overcome writer's block. By exploring new prompts and perspectives, writers can gain new insights and inspiration that can help them create compelling and original content.

Brainstorming Story Ideas

Brainstorming story ideas can be a challenging task for many writers. Here are some ways that ChatGPT can be used to generate story ideas:

1. Input a prompt: Input a general prompt such as "a character who discovers a hidden talent" and let ChatGPT generate story ideas based on that prompt. The generated ideas can be used to inspire and develop new storylines.
2. Character development: Input a character name and let ChatGPT generate details about the character's background, personality, and motivations. This can help writers develop more complex and interesting characters for their stories.
3. Setting exploration: Input a setting or location and let ChatGPT generate details about the

environment, culture, and history of that location. This can help writers create more immersive and detailed worlds for their stories.

4. Plot twists: Input a basic storyline and let ChatGPT generate potential plot twists and surprises that can add complexity and interest to the story.

5. Genre exploration: Input a genre such as "mystery" or "science fiction" and let ChatGPT generate potential story ideas within that genre. This can help writers explore new genres and develop new storylines.

Overall, ChatGPT can be a valuable tool for writers looking to brainstorm new story ideas and overcome writer's block. By generating prompts, characters, settings, and plot twists, writers can gain new perspectives and inspiration that can help them create engaging and original stories.

Character and Setting Development

ChatGPT can be a helpful tool for writers looking to develop characters and settings for their stories. Here are some ways that ChatGPT can be used for character and setting development:

1. Input character traits: Input a list of character traits such as "brave, loyal, and ambitious" and let ChatGPT generate a character based on those traits. This can help writers develop more complex and interesting characters for their stories.

2. Historical research: Input a historical event or time period and let ChatGPT generate details about the culture, politics, and environment of that period. This can help writers create more immersive and accurate settings for their stories.

3. Geographical research: Input a location or region and let ChatGPT generate details about the geography, climate, and natural resources of that area. This can help writers create more detailed and realistic settings for their stories.

4. Input a backstory: Input a character's backstory or history and let ChatGPT generate details about the character's upbringing, family, and experiences. This can help writers create more well-rounded and relatable characters for their stories.

5. Dialogue generation: Input a character's personality traits and let ChatGPT generate dialogue that reflects those traits. This can help writers develop more authentic and distinct voices for their characters.

Overall, ChatGPT can be a valuable tool for writers looking to develop more interesting and immersive characters and settings for their stories. By inputting different traits, events, and historical periods, writers can generate new perspectives and ideas that can help them create compelling and original fiction.

Chapter 3:
Plot Development and Story Arcs

ChatGPT can also be a helpful tool for writers looking to develop their story plots and story arcs. Here are some ways that ChatGPT can be used for plot development and story arcs:

1. Input a basic storyline: Input a basic storyline or plot summary and let ChatGPT generate potential plot points and twists that can add complexity and interest to the story.
2. Conflict generation: Input a character or setting and let ChatGPT generate potential conflicts that can drive the story forward. This can help writers develop more engaging and dynamic storylines.
3. Input a theme: Input a theme or message that the story is exploring and let ChatGPT generate potential plot points and conflicts that can develop and explore that theme. This can help writers create more cohesive and meaningful story arcs.
4. Genre exploration: Input a genre such as "thriller" or "romance" and let ChatGPT generate potential plot points and twists within that genre. This can help writers develop more compelling and appropriate storylines for their chosen genre.
5. Plot hole identification: Input a story draft and let ChatGPT identify potential plot holes or inconsistencies in the storyline. This can help writers identify areas where they need to develop the plot further or make adjustments to the existing storyline.

Overall, ChatGPT can be a useful tool for writers looking to develop more interesting and engaging story plots and story arcs. By inputting different elements such as storylines, themes, and genres, writers can generate new ideas and perspectives that can help them create more compelling and original fiction.

Crafting Compelling Plotlines with ChatGPT

Crafting compelling plotlines is an essential part of storytelling. ChatGPT can be a useful tool for writers looking to generate fresh ideas and overcome writer's block. Here are some ways to use ChatGPT to craft compelling plotlines:

1. Input a theme: Input a theme or topic that you want to explore in your story, and let ChatGPT generate potential plotlines that revolve around it. This can help you create a more focused and cohesive story.
2. Input a character: Input a character and let ChatGPT generate potential plotlines based on their

personality, backstory, and motivations. This can help you create a story that is more character-driven and emotionally resonant.

3. Input a genre: Input a genre such as science fiction, romance, or mystery, and let ChatGPT generate potential plotlines within that genre. This can help you develop a more engaging and appropriate story that fits the expectations of the genre.
4. Input a conflict: Input a conflict or challenge that the protagonist must face, and let ChatGPT generate potential plotlines that revolve around resolving that conflict. This can help you create a more focused and purposeful story.
5. Input a setting: Input a setting or location and let ChatGPT generate potential plotlines based on the environment, culture, and history of that place. This can help you create a more immersive and authentic story.

Overall, ChatGPT can be a helpful tool for writers looking to develop fresh and compelling plotlines. By inputting different elements such as themes, characters, genres, conflicts, and settings, writers can generate new ideas and perspectives that can help them create more engaging and original stories.

Building Conflict and Resolution

Conflict and resolution are key elements of storytelling. Here are some ways to use ChatGPT to build conflict and resolution in your writing:

1. Input a character: Input a character and let ChatGPT generate potential conflicts based on their personality, backstory, and motivations. This can help you create more interesting and complex conflicts that are grounded in the character's inner struggles.
2. Input a setting: Input a setting or location and let ChatGPT generate potential conflicts based on the environment, culture, and history of that place. This can help you create conflicts that are unique to the setting and feel authentic and immersive.
3. Input a theme: Input a theme or message that the story is exploring and let ChatGPT generate potential conflicts that can help develop and explore that theme. This can help you create more meaningful and impactful conflicts that serve the overall message of the story.
4. Input a genre: Input a genre such as thriller or romance and let ChatGPT generate potential conflicts that are appropriate and engaging for that genre. This can help you create conflicts that are compelling and keep readers invested in the story.
5. Input a resolution: Input a conflict or challenge and let ChatGPT generate potential resolutions or outcomes. This can help you develop satisfying and meaningful resolutions that tie up the story in a way that feels satisfying to readers.

Overall, ChatGPT can be a valuable tool for building conflict and resolution in your writing. By inputting different elements such as characters, settings, themes, genres, and resolutions, you can generate fresh ideas and perspectives that can help you create more interesting and engaging conflicts and resolutions.

Chapter 4:
Writing Dialogue with ChatGPT

Writing dialogue is an important part of storytelling, and ChatGPT can be a helpful tool for generating fresh and realistic dialogue. Here are some ways to use ChatGPT to write dialogue:

1. Input a character: Input a character and let ChatGPT generate potential dialogue based on their personality, tone, and voice. This can help you create dialogue that is authentic and unique to the character.
2. Input a scene: Input a scene or situation and let ChatGPT generate potential dialogue that fits the tone and mood of the scene. This can help you create dialogue that is appropriate and engaging for the context.
3. Input a conflict: Input a conflict or tension between characters and let ChatGPT generate potential dialogue that reflects the conflict and builds tension. This can help you create dialogue that is emotionally charged and moves the story forward.
4. Input a genre: Input a genre such as comedy or drama and let ChatGPT generate potential dialogue that is appropriate for that genre. This can help you create dialogue that is engaging and entertaining for readers.
5. Input a goal: Input a character's goal or objective and let ChatGPT generate potential dialogue that reflects the character's motivations and drives the plot forward. This can help you create dialogue that is purposeful and moves the story in a specific direction.

Overall, ChatGPT can be a useful tool for generating dialogue that is authentic, engaging, and appropriate for the context of the story. By inputting different elements such as characters, scenes, conflicts, genres, and goals, writers can generate new ideas and perspectives that can help them create more interesting and effective dialogue.

Creating Authentic Dialogue

Creating authentic dialogue is crucial for making characters feel real and engaging readers in a story. Here are some tips for using ChatGPT to create authentic dialogue:

1. Input realistic language: Use realistic language that matches the characters' personalities, backgrounds, and experiences. ChatGPT can help with this by generating dialogue based on the input you provide.

2. Use contractions: People use contractions in everyday speech, so incorporating them into your dialogue can make it feel more natural and authentic.
3. Vary sentence structure: People don't speak in perfectly constructed sentences, so varying sentence structure and using fragments can help make your dialogue feel more authentic.
4. Avoid info-dumping: People don't typically give long explanations or monologues in everyday conversation, so avoid info-dumping through dialogue. Instead, find ways to convey information through actions or other means.
5. Use subtext: People often communicate more through what they don't say than what they do say. Incorporating subtext into your dialogue can make it more complex and realistic.
6. Read it aloud: Reading your dialogue aloud can help you identify areas that sound unnatural or stilted. ChatGPT can generate dialogue, but it's up to you to refine it and make it sound authentic.

Overall, using ChatGPT to generate dialogue is a great way to jumpstart your writing and get new ideas, but it's important to remember that the dialogue you generate still needs to be refined and polished to feel truly authentic. By incorporating these tips and taking the time to refine your dialogue, you can create characters who feel real and engage readers in your story.

Improving Character Interaction

Improving character interaction is important for creating engaging and believable relationships in your writing. Here are some tips for using ChatGPT to improve character interaction:

1. Input multiple characters: Inputting multiple characters into ChatGPT can help you generate dialogue that showcases their personalities and builds relationships between them.
2. Use nonverbal cues: People communicate through nonverbal cues like body language and tone of voice, so incorporating these into your dialogue can help make character interactions feel more realistic.
3. Vary dialogue tags: Using a variety of dialogue tags like "said," "whispered," and "shouted" can help convey the tone of a character's voice and create more interesting interactions.
4. Build tension: Tension between characters can create engaging conflict and make character interactions more interesting. ChatGPT can help you generate dialogue that builds tension and reveals underlying conflicts.
5. Show, don't tell: Instead of telling readers about characters' relationships, show them through their interactions. ChatGPT can help you generate dialogue that shows characters' relationships rather than telling readers about them.

6. Consider power dynamics: Power dynamics between characters can affect their interactions. ChatGPT can help you generate dialogue that reflects power dynamics and builds tension between characters.

Overall, ChatGPT can be a helpful tool for generating dialogue that improves character interaction. By inputting multiple characters, using nonverbal cues, varying dialogue tags, building tension, showing rather than telling, and considering power dynamics, you can use ChatGPT to create more engaging and believable character interactions in your writing.

Chapter 5:
Revision and Editing with ChatGPT

Revision and editing are crucial steps in the writing process that help improve the clarity and quality of your work. While ChatGPT is a useful tool for generating ideas and content, it is important to use other methods for revision and editing. Here are some ways you can use ChatGPT for revision and editing:

1. Generate alternative phrasing: ChatGPT can help you generate alternative phrasing for sentences or paragraphs that are unclear or awkwardly worded.
2. Check for grammar and spelling errors: ChatGPT can help identify grammar and spelling errors in your writing, although it may not catch every mistake.
3. Get ideas for rephrasing: ChatGPT can provide suggestions for rephrasing sentences or paragraphs that are too repetitive or need to be reworded for clarity.
4. Identify inconsistencies: ChatGPT can help identify inconsistencies in your writing, such as character traits that change throughout the story or plot points that don't make sense.
5. Improve pacing: ChatGPT can help you identify areas where the pacing of your writing may be too slow or too fast, and suggest ways to improve it.
6. Use as a writing prompt: If you're stuck on a particular scene or chapter, ChatGPT can provide writing prompts to help you get unstuck and generate new ideas.

Overall, while ChatGPT can be a helpful tool for revision and editing, it is important to use it in conjunction with other methods of revision and editing, such as reading your work aloud, getting feedback from others, and using grammar and spell-checking software. By using a variety of tools and methods, you can improve the quality of your writing and ensure that it is clear, engaging, and error-free.

Proofreading for Grammar and Punctuation

Proofreading for grammar and punctuation is an important part of the writing process that helps ensure your writing is clear, effective, and error-free. While ChatGPT can help identify some grammar and punctuation errors, it is important to use other methods as well. Here are some tips for proofreading for grammar and punctuation using ChatGPT:

1. Input individual sentences: Inputting individual sentences into ChatGPT can help identify grammar and punctuation errors, such as subject-verb agreement or misplaced commas.

2. Use grammar-checking software: While ChatGPT can help identify some grammar and punctuation errors, dedicated grammar-checking software like Grammarly or ProWritingAid may be more effective at identifying specific types of errors.
3. Check for consistency: ChatGPT can help identify inconsistencies in your writing, such as using different tenses or styles of writing. Make sure your writing is consistent throughout.
4. Focus on one type of error at a time: It can be helpful to focus on one type of error, such as comma usage or subject-verb agreement, when proofreading. This can help you catch more errors and improve the overall quality of your writing.
5. Proofread multiple times: It is important to proofread your writing multiple times to catch all errors. Take breaks between proofreading sessions to avoid getting too tired or losing focus.
6. Get feedback from others: Ask someone else to read your writing and provide feedback on grammar and punctuation errors. A fresh set of eyes can often catch mistakes you may have missed.

Overall, while ChatGPT can be a useful tool for proofreading for grammar and punctuation errors, it is important to use it in conjunction with other methods and to proofread your writing multiple times to ensure it is error-free. By taking the time to proofread your writing and using a variety of tools and methods, you can improve the clarity and effectiveness of your writing.

Refining Style and Tone

Refining your style and tone is an important part of the writing process that helps ensure your writing is engaging, professional, and appropriate for your intended audience. While ChatGPT can offer suggestions for style and tone, here are some additional tips for refining your style and tone:

1. Know your audience: Before you begin writing, consider who your audience is and what they are looking for. Understanding your audience can help you choose the appropriate tone and style for your writing.
2. Choose a consistent style: Consistency is key when it comes to style. Choose a style that suits your audience and stick with it throughout your writing.
3. Vary your sentence structure: Varying your sentence structure can help keep your writing interesting and engaging. Mix up short and long sentences, and experiment with using different types of punctuation.
4. Use active voice: Active voice can make your writing more engaging and easier to read. Try to use active voice whenever possible.
5. Avoid jargon and overly technical language: Unless you are writing for a highly specialized

audience, it is best to avoid using jargon and overly technical language. Use simple, clear language that is easy for your audience to understand.

6. Use strong verbs and adjectives: Using strong verbs and adjectives can make your writing more engaging and descriptive. Avoid overusing adverbs, which can weaken your writing.
7. Get feedback from others: Ask someone else to read your writing and provide feedback on style and tone. They may be able to offer suggestions or identify areas where you can improve.

Overall, refining your style and tone takes time and effort. By understanding your audience, choosing a consistent style, varying your sentence structure, using active voice, avoiding jargon and overly technical language, using strong verbs and adjectives, and getting feedback from others, you can improve the overall quality of your writing and ensure it is engaging and appropriate for your intended audience.

Chapter 6:
Overcoming Writer's Block with ChatGPT

Writer's block is a common problem that many writers face. Fortunately, ChatGPT can be a helpful tool for overcoming writer's block. Here are some ways you can use ChatGPT to break through writer's block:

1. Generate new ideas: ChatGPT can be used to generate new ideas for your writing. Simply provide a prompt or topic, and ChatGPT can generate a list of ideas to help get you started.
2. Get unstuck: If you are stuck on a particular scene or plot point, ChatGPT can help you brainstorm new directions for your writing. Simply provide a description of the problem, and ChatGPT can offer suggestions for how to move forward.
3. Explore different perspectives: ChatGPT can be used to explore different perspectives on a topic or issue. By getting a fresh perspective, you may be able to overcome writer's block and approach your writing from a new angle.
4. Practice writing exercises: ChatGPT can generate writing prompts and exercises that can help you practice your writing and get your creative juices flowing.
5. Use as a sounding board: If you're struggling to get feedback on your writing, you can use ChatGPT as a sounding board. By providing a description of your work, ChatGPT can offer suggestions and feedback to help you improve your writing.

Overall, ChatGPT can be a useful tool for writers looking to overcome writer's block. By using it to generate new ideas, explore different perspectives, practice writing exercises, and get feedback on your work, you can break through writer's block and continue making progress on your writing.

ChatGPT as a Creativity Booster

ChatGPT can definitely serve as a creativity booster for writers. Here are a few ways in which it can help:

1. Brainstorming Ideas: ChatGPT can generate a large number of ideas and suggestions based on a given topic or prompt. This can help writers think outside the box and generate new and creative ideas that they may not have come up with on their own.
2. Providing Inspiration: ChatGPT can help provide writers with inspiration and new perspectives on a topic. By generating unique and interesting responses, it can help writers see things in a new light and spark their creativity.

3. Improving Writing Skills: ChatGPT can also help writers improve their writing skills by offering suggestions on grammar, syntax, and word choice. This can help writers develop their own writing style and learn new techniques to make their writing more creative and engaging.
4. Overcoming Writer's Block: As mentioned earlier, ChatGPT can be used to overcome writer's block by providing new ideas, perspectives, and directions for a writer's work. By using ChatGPT to generate prompts and exercises, writers can also break out of their routine and try new approaches to their writing.

Overall, ChatGPT can serve as a powerful tool for boosting creativity in writers. By providing new ideas, inspiration, and guidance, it can help writers expand their thinking and develop their skills, ultimately leading to more creative and engaging writing.

Keeping the Writing Process Flowing

Keeping the writing process flowing can be a challenge for many writers. Here are some ways ChatGPT can help keep the creative juices flowing:

1. Writing Prompts: ChatGPT can generate a variety of writing prompts to help writers get started and keep their writing flowing. These prompts can be specific or general, and can be tailored to the writer's preferences or needs.
2. Sentence Completion: ChatGPT can offer sentence completion suggestions to help writers overcome writer's block and keep their writing moving forward. This can be particularly helpful when a writer is struggling to come up with the next sentence or idea.
3. Idea Expansion: ChatGPT can also help writers expand on their existing ideas by offering suggestions on how to develop them further. This can include brainstorming new angles or perspectives, as well as suggesting new characters, settings, or plot twists.
4. Revision and Editing: ChatGPT can help writers keep their writing flowing by offering suggestions for revision and editing. By providing guidance on how to improve their work, writers can stay engaged and motivated throughout the writing process.

Overall, ChatGPT can be an invaluable tool for writers looking to keep their writing process flowing. By offering a variety of suggestions and ideas, it can help writers stay inspired and engaged, leading to more productive and creative writing sessions.

Chapter 7:
Genre-Specific Writing with ChatGPT

ChatGPT can be used to generate ideas and suggestions for a wide range of writing genres, including fiction, non-fiction, poetry, and more. Here are some ways ChatGPT can be used to help with genre-specific writing:

1. Fiction: ChatGPT can help writers with fiction by generating ideas for characters, settings, and plot twists. It can also offer suggestions for different genres of fiction, such as romance, sci-fi, or fantasy.
2. Non-fiction: ChatGPT can be used to help writers with non-fiction by generating ideas for topics, research, and organization. It can also offer suggestions on how to make complex ideas accessible to readers.
3. Poetry: ChatGPT can be used to help poets with their work by generating ideas for themes, imagery, and poetic forms. It can also offer suggestions on how to use language and structure to create powerful and evocative poetry.
4. Screenplays: ChatGPT can be used to help screenwriters by generating ideas for scenes, characters, and dialogue. It can also offer suggestions on how to structure a screenplay and build tension and conflict.

Overall, ChatGPT can be a valuable tool for writers in any genre. By generating ideas and suggestions tailored to specific genres, it can help writers expand their creativity and improve their writing in new and exciting ways.

Writing for Fiction, Nonfiction, and Poetry

Writing for different genres requires different approaches and techniques. Here are some tips for writing in three common genres: fiction, nonfiction, and poetry.

Fiction:

- Create compelling characters with interesting backstories and motivations.
- Build a well-crafted plot with a clear beginning, middle, and end.
- Use vivid descriptions to bring settings and scenes to life.
- Write realistic and engaging dialogue that reveals character and moves the plot forward.
- Use tension and conflict to keep readers engaged.

Nonfiction:

- Start with a clear thesis statement that summarizes the main point of the piece.
- Use research to support your argument or ideas.
- Organize your writing in a logical and clear way.
- Use examples, anecdotes, and stories to illustrate your points.
- Write in a style that is accessible and easy to understand.

Poetry:

- Experiment with different poetic forms, such as sonnets or haikus.
- Use imagery and metaphor to create vivid and evocative language.
- Pay attention to the rhythm and sound of your words.
- Explore themes and emotions through your poetry.
- Edit and revise your work carefully to create a polished final product.

In all three genres, it's important to write with clarity, purpose, and authenticity. With the help of ChatGPT, writers can generate new ideas and inspiration for their writing, and hone their skills in these different genres.

Adapting ChatGPT for Different Genres

ChatGPT can be adapted for different genres by adjusting the prompts and responses to fit the specific requirements of the genre. Here are some examples:

Fiction:

- Prompt: "Write a scene in which two characters confront each other about a secret they have been keeping from each other."
- Response: "As soon as John walked into the room, Mary knew she had to tell him the truth about what had happened the night before. She took a deep breath and began, 'John, I have to tell you something...'"

Nonfiction:

- Prompt: "Write an article about the benefits of meditation for mental health."
- Response: "Meditation has been shown to have numerous benefits for mental health, including reducing stress and anxiety, improving focus and concentration, and increasing feelings of calm and well-being."

Poetry:

- Prompt: "Write a poem about the changing of the seasons."
- Response: "The leaves fall gently to the ground As autumn breezes swirl around The air grows crisp and cool and clear As summer fades and winter draws near."

By adjusting the prompts and responses, writers can use ChatGPT as a tool for generating ideas and inspiration for their writing in different genres.

Chapter 8:
Writing for Different Mediums

ChatGPT can also be adapted for writing in different mediums, such as screenplays, stage plays, and video game scripts. Here are some examples:

Screenplays:

- Prompt: "Write a scene in which the main character confronts the villain."
- Response: "John storms into the villain's lair, determined to put an end to their evil plans. The villain sneers and taunts him, but John stands firm and delivers a powerful speech, outlining why their actions are wrong and how he intends to stop them."

Stage plays:

- Prompt: "Write a scene in which two characters have an argument that reveals secrets about their past."
- Response: "As they sit in the café, the tension between John and Mary is palpable. They begin to argue about a seemingly small matter, but as the argument continues, it becomes clear that there are deeper issues at play. Mary reveals a secret from John's past that he has been keeping hidden, and he lashes out in anger and frustration."

Video game scripts:

- Prompt: "Write a dialogue between the player character and an NPC (non-player character) who gives them a quest."
- Response: "As the player approaches the NPC, they notice a glint in their eye that suggests they have a task for them. The NPC explains that a valuable artifact has been stolen and asks the player to retrieve it. They give the player a clue as to its location and a warning that there may be dangerous enemies guarding it."

By adapting ChatGPT to different mediums, writers can use it to generate ideas and inspiration for their writing in a variety of formats.

Using ChatGPT for Screenwriting, Playwriting, and Blogging

ChatGPT can be a valuable tool for screenwriting, playwriting, and blogging. Here are some ways in which it can be used in each of these areas:

Screenwriting

- Idea generation: ChatGPT can be used to generate ideas for characters, plotlines, and scenes. For example, you can prompt it with a question like "What is a unique concept for a horror movie?" and use its response to spark your creativity.
- Dialogue generation: ChatGPT can also be used to generate dialogue for your script. You can prompt it with a character name and a general topic, such as "What would John say to Mary about their past?" and use its response as a starting point for your scene.
- Plot development: ChatGPT can help you develop your plot by providing suggestions for how to move your story forward. For example, you can prompt it with a question like "How can I create a twist ending for my movie?" and use its response as inspiration for your script.

Playwriting

- Character development: ChatGPT can help you develop your characters by generating ideas for their backstory, personality, and motivations. You can prompt it with a question like "What is an interesting flaw for my protagonist?" and use its response to add depth to your character.
- Dialogue generation: ChatGPT can also be used to generate dialogue for your play. You can prompt it with a character name and a general topic, such as "What would Jane say to her mother about her marriage?" and use its response as a starting point for your scene.
- Scene development: ChatGPT can help you develop your scenes by providing suggestions for how to create conflict and tension. For example, you can prompt it with a question like "How can I create a powerful moment in my play?" and use its response to inspire your scene.

Blogging:

- Topic generation: ChatGPT can be used to generate ideas for blog posts. You can prompt it with a question like "What are some unique topics for a food blog?" and use its response to spark your creativity.
- Content development: ChatGPT can help you develop your blog content by generating ideas for your post. For example, you can prompt it with a question like "What are some ways to save money on groceries?" and use its response to create your blog post.
- SEO optimization: ChatGPT can also be used to optimize your blog post for search engines. You can prompt it with a topic and a keyword, such as "How to make chocolate cake" and "chocolate cake recipe," and use its response to create a blog post that is optimized for search engines.

By using ChatGPT in these ways, writers can save time and generate ideas for their writing that they may not have thought of otherwise.

Tailoring Content to Specific Mediums

When writing for different mediums, it's important to tailor the content to fit the medium. For example, a blog post might have a more conversational tone and be more casual, while a play or screenplay might be more formal and have specific formatting requirements.

ChatGPT can be useful for adapting writing to different mediums by generating content that fits the tone and style required for the specific medium. For example, if you're writing a blog post and need a catchy introduction, ChatGPT can generate a variety of options for you to choose from. Similarly, if you're writing a screenplay and need a specific type of dialogue, ChatGPT can generate examples that fit the tone and style of the medium.

It's important to keep in mind that ChatGPT is a tool and should be used as such. It can generate ideas and content, but it's up to the writer to use their own judgement to determine what works best for the specific medium and the intended audience.

Chapter 9:
Ethical Considerations in AI-Assisted Writing

As with any technology, there are ethical considerations when it comes to using AI for writing. One of the main concerns is the potential for plagiarism. While AI-generated content can be a helpful tool for generating ideas and providing inspiration, it's important to ensure that the content being produced is original and not copied from existing sources. It's important to use AI-generated content as a starting point for your own writing, and not rely on it entirely.

Another concern is the potential for bias in AI-generated content. Just as with any AI technology, the data used to train ChatGPT can be biased, which can lead to biased output. It's important to be aware of this and to take steps to mitigate bias by using diverse training data and reviewing and editing AI-generated content to ensure it is free from any discriminatory language or ideas.

Lastly, it's important to consider the impact of AI on the job market for writers. While AI can be a helpful tool for writers, there is the potential for it to replace human writers in certain areas, which can lead to job losses. It's important to balance the benefits of AI with the potential impact on employment in the writing industry.

Plagiarism and Originality

Plagiarism and originality are important considerations when it comes to using AI for writing. While AI-generated content can be a helpful tool for generating ideas and providing inspiration, it's important to ensure that the content being produced is original and not copied from existing sources.

One way to ensure originality is to use AI-generated content as a starting point for your own writing, rather than relying on it entirely. This means using the content as a source of inspiration and using it to generate ideas, but then developing those ideas into original content that is uniquely your own.

Another way to ensure originality is to use plagiarism detection tools to check your writing against existing sources. This can help you identify any instances of unintentional plagiarism and make sure that your writing is entirely original.

It's also important to be aware of copyright laws and to ensure that any content you produce using AI-generated content is legally and ethically sound. This means being careful not to infringe on the intellectual property rights of others and to make sure that any content you produce is original and belongs to you.

Overall, while AI-generated content can be a helpful tool for writing, it's important to ensure that any content produced is original and does not infringe on the intellectual property rights of others.

Responsible Use of AI in Writing

Responsible use of AI in writing involves several considerations, including ethical and legal considerations, as well as ensuring the quality and accuracy of the content produced.

Firstly, it's important to ensure that any content produced using AI is accurate and of high quality. This means being selective in the sources of data and algorithms used to generate the content, and verifying the accuracy of the content produced.

Secondly, it's important to be aware of any legal and ethical implications of using AI-generated content. For example, using AI to create content that infringes on intellectual property rights or violates privacy laws can have serious legal consequences. It's important to ensure that any content produced using AI is legal and ethically sound.

Another consideration is ensuring that the content produced is appropriate and relevant for the intended audience. This means understanding the audience and tailoring the content to their needs and interests.

Finally, it's important to be transparent about the use of AI in creating content. This means being upfront about the use of AI and ensuring that readers and other stakeholders are aware of the role that AI played in the creation of the content.

Overall, responsible use of AI in writing involves ensuring the accuracy and quality of the content produced, being aware of any legal and ethical implications, tailoring the content to the intended audience, and being transparent about the use of AI.

Chapter 10:
Looking Ahead: ChatGPT and the Future of Writing

The future of writing with AI and ChatGPT is promising, with potential applications in a variety of fields such as journalism, marketing, and content creation. As AI continues to improve, it has the potential to take on more sophisticated writing tasks and become an even more powerful tool for writers.

One potential area of development is in the use of AI for personalized writing. As AI becomes better at understanding individual users and their preferences, it may be possible to use it to generate highly customized content that is tailored to specific individuals or groups.

Another area of potential growth is in the use of AI for multilingual writing. As AI improves its ability to understand and translate between languages, it may become possible to generate content in multiple languages simultaneously, making it easier for writers to reach global audiences.

There are also opportunities for AI to help address some of the challenges facing the writing industry, such as the need for increased productivity and the demand for high-quality content. By automating some of the more time-consuming and tedious aspects of writing, such as proofreading and editing, writers may be able to focus more on creating high-quality content.

Overall, the future of writing with AI and ChatGPT is exciting, and there are many opportunities for writers to take advantage of these technologies to improve their work and reach new audiences. As AI continues to develop and become more advanced, it will likely become an even more powerful tool for writers and creators.

Current Trends and Future Predictions

One of the current trends in the use of ChatGPT for writing is the development of more sophisticated and specific models tailored to particular genres and mediums. This allows for greater accuracy and flexibility in generating content for different contexts.

Another trend is the increasing use of AI in content creation and marketing, with companies using AI-generated content to engage with audiences and drive conversions.

In the future, it is likely that AI will become an even more integral part of the writing process, with writers using it for everything from idea generation to editing and proofreading. As AI continues to improve, it may become possible for it to take on more complex writing tasks, such as creating entire novels or screenplays.

However, there are also concerns about the potential impact of AI on the writing industry, particularly in terms of job displacement and the potential for AI-generated content to be less original or creative than content generated by human writers.

Overall, the use of ChatGPT and AI in writing is likely to continue to grow and evolve in the coming years, with both benefits and challenges for the industry and the writers who work within it.

Embracing AI in the Writing Process

As the use of AI in writing continues to grow, it's important for writers to consider how they can effectively incorporate these tools into their creative process. Here are some tips for embracing AI in the writing process:

1. Understand the capabilities and limitations of AI: Before using AI for writing, it's important to have a good understanding of what it can and can't do. While AI can be helpful in generating ideas and improving efficiency, it's not a replacement for human creativity and intuition.
2. Use AI for specific tasks: Instead of relying on AI for the entire writing process, consider using it for specific tasks, such as generating ideas, conducting research, or proofreading. This allows you to capitalize on the strengths of AI while still maintaining creative control over the final product.
3. Collaborate with AI: Think of AI as a tool to collaborate with, rather than a replacement for human creativity. By working with AI to generate ideas or refine drafts, you can take advantage of its capabilities while still maintaining your unique voice and perspective as a writer.
4. Experiment and adapt: As with any new technology, it's important to experiment with AI and find the best ways to integrate it into your writing process. Be open to trying new tools and techniques, and be willing to adapt as you learn more about what works best for you.

Overall, the key to successfully embracing AI in the writing process is to approach it as a complementary tool, rather than a replacement for human creativity and intuition. By leveraging the strengths of AI while still maintaining creative control over the final product, writers can unlock new levels of efficiency and innovation in their work.

BOOK 6: "ChatGPT for Programmers: Code, Debugging and Optimization": Discusses the use of ChatGPT as a programming assistant.

Chapter 1:
Introduction to ChatGPT for Programmers

Welcome to the Introduction to ChatGPT for Programmers. In this discussion, we will cover the basics of ChatGPT, its applications in programming, and how programmers can use it to enhance their work.

Understanding ChatGPT

ChatGPT (Generative Pre-trained Transformer 3) is an advanced AI language model developed by OpenAI. It uses machine learning algorithms to understand human language and generate responses that are almost indistinguishable from those of human beings.

ChatGPT is pre-trained on massive amounts of data, making it capable of generating coherent and contextually relevant responses to a wide range of questions and statements. This allows it to be used for various applications, including natural language processing, text generation, and conversation generation.

ChatGPT is built on the Transformer architecture, which allows it to perform language tasks with a high level of accuracy and efficiency. It has become increasingly popular in recent years due to its ability to generate human-like responses to a wide range of inputs.

The Role of AI in Programming

AI has several roles in programming. One of the most significant roles of AI in programming is automating repetitive tasks, such as debugging, testing, and deployment. By using AI algorithms, developers can reduce the time it takes to test and deploy software applications, which can lead to faster development cycles.

AI is also used in programming for data analysis and optimization. Developers can use machine learning algorithms to analyze large data sets and identify patterns and trends that are not immediately apparent to humans. This information can be used to optimize applications and improve user experiences.

Additionally, AI is used in programming for natural language processing and speech recognition. Developers can use AI algorithms to build chatbots, virtual assistants, and other conversational interfaces that can communicate with users in a natural and intuitive way. This can improve customer engagement and lead to more efficient and effective communication.

Overall, AI has the potential to revolutionize the programming industry by enabling developers to automate repetitive tasks, optimize applications, and create more natural and intuitive interfaces for users.

Chapter 2:
Code Generation with ChatGPT

Code generation with ChatGPT involves using AI algorithms to automate the process of writing code. With this approach, developers can input a high-level description of the code they need, and the AI will generate the code for them.

One way that ChatGPT can assist with code generation is by helping developers write more efficient and effective algorithms. By inputting data sets and other parameters, ChatGPT can analyze the data and generate code that is optimized for the specific task at hand.

Another way that ChatGPT can assist with code generation is by automating the process of writing boilerplate code. Developers can input a description of the application they are building, and ChatGPT can generate the code for the basic structure of the application. This can save developers time and help them focus on more complex programming tasks.

ChatGPT can also assist with the creation of APIs and libraries by generating code that conforms to industry standards and best practices. This can help developers build more robust and interoperable applications.

Overall, code generation with ChatGPT can significantly reduce the time and effort required to write code, while also improving code quality and consistency. However, it is important for developers to carefully review the code generated by AI algorithms to ensure that it is correct and meets their specific requirements.

Writing Code with AI Assistance

Writing code with the help of AI can significantly speed up the development process and improve code quality. ChatGPT can be used to generate code in various programming languages, including Python, JavaScript, and C++.

One way to use ChatGPT for code generation is to provide it with a task description or a set of requirements and let it generate the code based on that. For example, if you want to create a function that finds the average of a list of numbers in Python, you could provide the following prompt to ChatGPT: "Write a Python function that takes a list of numbers as input and returns the average of those numbers." ChatGPT would then generate the code for the function based on its understanding of Python syntax and the requirements you provided.

Another way to use ChatGPT for code generation is to provide it with a sample of existing code and let it generate similar code. This can be particularly useful when you need to write repetitive code, such as boilerplate code for a new project. For example, if you have a piece of Python code that reads data from a CSV file and creates a list of dictionaries, you could provide that code to ChatGPT and ask it to generate similar code that reads data from a different file format, such as JSON.

It's important to note that the code generated by ChatGPT may not always be optimal or even correct. As with any AI system, ChatGPT's output should always be carefully reviewed and tested before it's used in production. Additionally, the use of AI-generated code raises ethical concerns around ownership and originality, which should be carefully considered.

Generating Boilerplate and Template Code

Boilerplate and template code refer to pieces of code that are used frequently and repeatedly in software development. Boilerplate code is often necessary but does not directly contribute to the functionality of the software, while template code provides a basic structure for the software to build on.

ChatGPT can be used to generate boilerplate and template code quickly and easily. For example, a programmer can input a few parameters and requirements for a specific piece of code, and ChatGPT can generate a complete template for that code. This can save time and effort, especially when working on large projects with a lot of repetitive code. Additionally, ChatGPT can generate code in different programming languages, allowing programmers to work on projects in languages they may not be familiar with.

Chapter 3:
Debugging with ChatGPT

Debugging is the process of identifying and fixing errors, bugs, and other issues in software code. It can be a time-consuming and challenging task, especially when dealing with complex software systems. ChatGPT can assist in debugging by generating suggestions for possible solutions to the errors identified.

By inputting the error message or problem statement, ChatGPT can provide possible causes and solutions to the problem. Additionally, ChatGPT can analyze the code and identify potential areas of concern that may be causing the error. This can save programmers significant time and effort, especially when dealing with complicated code.

It is important to note that ChatGPT's suggestions should always be verified and tested by the programmer to ensure that they are valid and effective. ChatGPT's suggestions are just that, suggestions, and should not be blindly implemented without thorough analysis and consideration.

Identifying Bugs with AI

ChatGPT can assist in identifying bugs in programming. By using natural language processing and machine learning techniques, it can analyze the code and provide suggestions for fixing bugs. For example, it can analyze error messages and suggest possible solutions or identify code patterns that are known to cause bugs.

Moreover, ChatGPT can also learn from the code and detect recurring errors or patterns that lead to bugs. It can also identify coding practices that can lead to potential bugs in the future, allowing programmers to fix the code before it causes issues.

Overall, ChatGPT can be a valuable tool for debugging and improving the quality of code. However, it should not replace the programmer's expertise and judgment, and the output should always be carefully reviewed and tested.

Suggesting Fixes and Improvements

Yes, ChatGPT can suggest fixes and improvements for code. One way it can do this is by using machine learning algorithms to analyze large datasets of code and identify patterns and common errors. Based on this analysis, it can make suggestions for code changes that are likely to be effective. Additionally, it can learn from the corrections made by developers and adjust its suggestions accordingly.

Another approach is to use natural language processing techniques to analyze code documentation and provide suggestions for improving code readability and maintainability. This can include suggestions for naming conventions, comments, and formatting.

Overall, ChatGPT has the potential to assist programmers in identifying and fixing bugs, improving code quality, and increasing productivity. However, it is important to note that its suggestions should be reviewed and validated by human programmers, as they may not always be accurate or appropriate for the specific context.

Chapter 4:
Code Review and Quality Assurance

Code review and quality assurance are important steps in the software development process to ensure that the code is efficient, secure, and reliable. AI can assist programmers in identifying and fixing code errors, improving code quality, and ensuring code adherence to industry standards.

ChatGPT can help programmers in code review and quality assurance in the following ways:

1. Automated code review: ChatGPT can perform automated code review to identify code errors, vulnerabilities, and other potential issues. It can check code for syntax errors, style violations, and adherence to best practices and standards.
2. Code optimization: ChatGPT can suggest code optimization techniques to improve code performance, reduce resource usage, and increase efficiency. It can also provide suggestions for code refactoring to make it more modular, maintainable, and scalable.
3. Test case generation: ChatGPT can generate test cases to validate code functionality and identify potential issues. It can also help in creating test cases for edge cases and other scenarios that may not have been considered by the programmer.
4. Security analysis: ChatGPT can assist in security analysis to identify potential vulnerabilities and suggest fixes. It can perform static code analysis to identify security vulnerabilities and recommend security best practices to ensure code security.
5. Code documentation: ChatGPT can assist in creating code documentation to make the code more understandable and maintainable. It can generate documentation for code functions, classes, and modules, making it easier for programmers to understand how the code works.

By leveraging AI to assist in code review and quality assurance, programmers can reduce the time and effort required for manual code review, enhance the quality of code, and improve the overall efficiency of the software development process.

Using ChatGPT for Code Review

ChatGPT can be used for code review by analyzing the code and providing suggestions for improvement based on existing best practices and standards. Here are some ways in which ChatGPT can be used for code review:

1. Code style review: ChatGPT can be used to analyze the code and provide suggestions for

improving the code style, such as code indentation, formatting, naming conventions, and documentation.

2. Code quality review: ChatGPT can be used to analyze the code and provide suggestions for improving code quality, such as identifying potential bugs, inefficiencies, and security vulnerabilities.
3. Code review automation: ChatGPT can be used to automate the code review process by analyzing the code and comparing it to existing code review guidelines and standards. This can help reduce the time and effort required for manual code review.
4. Code review collaboration: ChatGPT can be used to facilitate collaboration between developers by providing a platform for sharing and reviewing code. Developers can use ChatGPT to provide feedback on code changes and collaborate on improving code quality.

Overall, ChatGPT can be a valuable tool for improving code quality and streamlining the code review process in software development.

Ensuring Code Quality with AI

Artificial intelligence (AI) can be used to improve code quality by detecting potential issues before the code is deployed. AI-powered tools can analyze code to identify errors, security vulnerabilities, performance issues, and more. By catching these problems early on, developers can fix them before they cause problems for users.

Some ways in which ChatGPT can be used for code review and quality assurance are:

1. Automated code reviews: AI-powered tools can analyze code automatically and provide feedback on issues such as syntax errors, style violations, and security vulnerabilities.
2. Predictive analytics: AI can be used to identify patterns and predict potential problems in code. This can help developers to fix issues before they cause problems for users.
3. Code optimization: AI can be used to optimize code for performance, memory usage, and other factors. This can help to improve the overall quality of the code.
4. Test automation: AI-powered tools can automate testing, making it faster and more accurate. This can help to ensure that code is functioning as intended.
5. Code suggestion: AI can suggest code snippets and provide code completion suggestions that can help to speed up the development process and ensure that code is accurate and error-free.

Overall, using ChatGPT for code review and quality assurance can help to improve the overall quality of code, reduce development time, and improve user experience.

Chapter 5:
Automated Testing with ChatGPT

Automated testing is the process of testing software or applications using software tools to ensure that it performs as expected and meets the requirements. ChatGPT can be used to automate the testing process by generating test cases and scripts.

ChatGPT can analyze the code and identify potential bugs and errors. It can also generate test cases that cover all possible scenarios and edge cases, which can help ensure that the software is thoroughly tested.

Additionally, ChatGPT can be used to automate the execution of test cases, reducing the amount of manual effort required to run tests. This can help improve efficiency and speed up the testing process.

Overall, ChatGPT can be a valuable tool in the software testing process, helping to improve the quality of software and reduce the risk of errors and bugs.

Writing Test Cases with AI Assistance

AI can assist programmers in writing test cases for software applications. Test cases are written to ensure that the application meets the desired functional requirements and to identify any defects or bugs. ChatGPT can assist in writing test cases by suggesting inputs and expected outputs for different functions or methods. It can also help in generating test data that covers a wide range of input scenarios.

ChatGPT can analyze the code and provide suggestions for testing different parts of the code. It can also identify any potential edge cases or scenarios that may need additional testing. Additionally, ChatGPT can help in generating automated test scripts that can be used to run tests repeatedly and ensure consistent results.

By using ChatGPT for test case generation and automated testing, programmers can save time and effort in writing and executing test cases, while also ensuring the quality of the software application.

Automating Test Execution and Reporting

ChatGPT can be utilized for automating test execution and reporting. Test automation refers to the process of using software tools to control the execution of tests and the comparison of actual outcomes with predicted outcomes. The primary goal of test automation is to ensure that the software product meets the expected quality standards, and it is a crucial step in the software development process.

ChatGPT can be used for automating the test execution process by generating test cases based on the requirements and specifications. Test cases can be generated automatically, and ChatGPT can be trained to recognize patterns and generate test cases based on the input data. This can significantly reduce the time and effort required for manual test case generation.

ChatGPT can also be used for test reporting, which is the process of documenting and communicating the results of the testing process. Test reporting involves generating reports that summarize the test results, identify any issues, and provide recommendations for improvement.

ChatGPT can be used to generate test reports automatically, which can save time and improve the accuracy of the reporting process. ChatGPT can be trained to analyze the test results and identify any issues, and it can generate reports that summarize the results and provide recommendations for improvement. This can help organizations to identify and address any issues in a timely and efficient manner, which can improve the overall quality of the software product.

Chapter 6:
Documentation and Comments

Documentation and comments are important aspects of programming that allow developers to understand and maintain code. Documentation provides information about the code's functionality, purpose, and usage, while comments explain the code's logic and provide context. AI can be used to assist with both documentation and commenting.

One way AI can help with documentation is by generating documentation automatically based on the code. Natural language processing models can analyze the code and generate documentation that explains the code's purpose, input parameters, return values, and usage. This can save developers time and ensure that code is well-documented.

AI can also help with commenting by analyzing the code and generating comments that explain the code's logic. This can be particularly useful for complex algorithms or code that is difficult to understand. AI-generated comments can provide context and clarity to the code, making it easier to read and maintain.

However, it is important to note that AI-generated comments and documentation may not always be accurate or complete. Developers should review and verify the generated documentation and comments to ensure that they are correct and provide the necessary information. Additionally, developers should still be encouraged to write their own documentation and comments to provide additional context and insights.

Generating Code Documentation with ChatGPT

One of the useful applications of ChatGPT in programming is generating code documentation. Documentation is an essential aspect of software development as it helps other developers understand how a program works and how to use its functions. With ChatGPT, programmers can generate code documentation automatically, saving them time and effort.

ChatGPT can analyze the code and generate comments that describe the code's purpose and functionality. It can also generate documentation for classes, functions, and variables, including information on their input and output parameters. By generating documentation automatically, programmers can ensure that their code is well-documented and easier to maintain, even if they don't have the time or resources to write documentation manually.

In addition to generating code documentation, ChatGPT can also help programmers create code comments. Code comments are used to explain what specific code blocks do or to provide context for other developers working on the codebase. ChatGPT can analyze the code and generate comments that explain how the code works, what it does, and why it's necessary.

Overall, using ChatGPT for code documentation and commenting can help improve the quality of code and make it more accessible to other developers.

Writing Effective Comments and Docstrings

Writing effective comments and docstrings is an essential aspect of programming. It helps other developers understand your code and makes maintenance and debugging easier. Here are some ways ChatGPT can assist in this area:

1. Writing clear and concise comments: You can use ChatGPT to help you write clear and concise comments that explain what your code is doing. Simply provide ChatGPT with a brief description of the code, and it can generate a comment that accurately describes the code.
2. Generating docstrings: Docstrings are an essential part of Python programming. They describe what a function does and what arguments it takes. ChatGPT can help you generate docstrings automatically by analyzing the function's arguments and return values.
3. Documenting code flow: Sometimes, it can be challenging to understand the flow of the code. In such cases, ChatGPT can help you generate comments that describe the logic of the code and how it flows.
4. Generating inline comments: You can also use ChatGPT to generate inline comments that explain a particular line or block of code. This can be especially useful if you're working with a large codebase or if you're collaborating with other developers.
5. Checking grammar and spelling: ChatGPT can also assist you in checking the grammar and spelling of your comments and docstrings, ensuring that they are clear and easy to understand.

In summary, ChatGPT can assist in generating clear and concise comments, generating docstrings, documenting code flow, generating inline comments, and checking grammar and spelling.

Chapter 7:
Optimization and Performance Tuning

Optimization and performance tuning are critical components of software development. Writing code that executes efficiently is essential to the success of any application, whether it is a desktop application, web application, or mobile app. AI can be used to help optimize and tune performance by analyzing code and identifying areas for improvement.

ChatGPT can assist with performance tuning and optimization by analyzing code and providing suggestions for improving performance. For example, it can identify code that is taking longer to execute than it should and recommend ways to optimize it. Additionally, it can suggest changes to algorithms or data structures that may improve performance.

ChatGPT can also help with memory management, which is an essential aspect of performance optimization. It can analyze code and identify areas where memory usage can be reduced, such as by reusing objects or reducing the size of data structures.

Overall, ChatGPT can assist with optimization and performance tuning by providing developers with insights and suggestions for improving the efficiency of their code.

Identifying Performance Bottlenecks with AI

AI can help identify performance bottlenecks in code by analyzing its execution patterns and identifying areas of code that are taking longer to execute than others. For example, AI can analyze the execution time of different functions in a program and identify which ones are taking the longest to execute. Once the bottlenecks are identified, developers can optimize the code to improve its performance.

AI can also help with performance tuning by suggesting code changes that can improve performance. For example, AI can suggest changes to algorithms, data structures, or code patterns that can improve performance. Additionally, AI can help with tuning system parameters such as memory allocation, thread management, and network settings to optimize the performance of the system.

Suggesting Optimization Strategies

ChatGPT can suggest optimization strategies that can be used to improve the performance of code. For example, it can analyze the code and identify inefficient algorithms or data structures that are causing performance issues. It can also recommend ways to parallelize the code or optimize memory usage to speed up execution. Additionally, ChatGPT can suggest ways to optimize code for specific hardware or platform, such as using vectorization or GPU acceleration. Overall, ChatGPT can be a valuable tool for programmers looking to improve the performance of their code.

Chapter 8:
ChatGPT for Learning New Programming Languages

ChatGPT can be a useful tool to learn new programming languages. Here are some ways in which it can help:

1. **Translation:** ChatGPT can translate code written in one programming language to another language. This can be helpful in learning a new programming language since you can see how code in one language is written in another language.
2. **Examples and Tutorials:** ChatGPT can generate examples and tutorials for different programming languages. These examples can help you understand the syntax and structure of a new programming language.
3. **Debugging:** ChatGPT can help in identifying and fixing errors in code written in a new programming language. You can input the code and ChatGPT will suggest fixes for errors.
4. **Project Ideas:** ChatGPT can suggest project ideas for a new programming language. This can be useful in practicing and implementing what you have learned.
5. **Answering Questions:** You can ask ChatGPT questions related to a new programming language and it can provide you with answers. This can help you in understanding the concepts and principles of the new language.

Overall, ChatGPT can be a valuable resource for learning new programming languages. It can help in translating code, generating examples and tutorials, debugging, suggesting project ideas, and answering questions.

Using ChatGPT as a Learning Resource

As a programmer, you can use ChatGPT to learn new programming languages by asking questions about syntax, data structures, and programming concepts. ChatGPT can provide explanations and examples of code, making it easier to understand and apply new concepts. You can also use ChatGPT to get recommendations for learning resources such as online tutorials, documentation, and books. ChatGPT can suggest practice exercises and challenges to help you improve your skills in a new programming language. Additionally, you can use ChatGPT to get help with common errors and issues you may encounter while learning a new language.

Translating Code Between Different Languages

ChatGPT can assist with translating code between different programming languages. The process involves using natural language processing to convert the code in one programming language into text, and then generating equivalent code in the target programming language. While this approach may not always produce perfect translations, it can save significant amounts of time and effort that would otherwise be required to manually rewrite code. It can also help programmers become proficient in new programming languages more quickly by providing them with an automated way to translate code they are already familiar with into a new language.

Chapter 9:
Ethical Considerations in AI-Assisted Programming

As with any technology, there are ethical considerations to take into account when using AI in programming. Here are some key issues to consider:

1. Bias: Like all machine learning models, ChatGPT may have biases built into it based on the data it was trained on. Programmers should be aware of this potential bias and take steps to mitigate it.
2. Accountability: When AI is used to make decisions in programming, it can be difficult to determine who is accountable for those decisions. Programmers need to take responsibility for the output of the AI they use.
3. Privacy: ChatGPT may require access to sensitive information to function properly. Programmers need to take steps to protect user privacy and ensure that any data they collect is stored securely.
4. Transparency: The use of AI in programming can make it difficult to understand how decisions are being made. Programmers should strive to make the AI they use as transparent as possible, so that the reasoning behind decisions can be understood.
5. Job displacement: As AI becomes more capable, it may displace jobs in the programming industry. It is important for programmers to be aware of this potential impact and take steps to mitigate it.

Responsible Use of AI in Coding

The responsible use of AI in coding involves being mindful of the potential consequences of using AI and taking steps to minimize any negative impacts. One important consideration is to ensure that AI is being used to enhance human decision-making, rather than replace it entirely. It's also important to be aware of any biases that may be present in the AI system and to take steps to address them. Additionally, developers should be transparent about their use of AI and the data they are using to train their models, as well as obtain informed consent from users when appropriate. Finally, developers should also be aware of any legal or regulatory frameworks that may apply to the use of AI in coding and ensure compliance with these regulations.

Addressing Bias and Fairness in AI Tools

Addressing bias and fairness in AI tools is essential to ensure that the outcomes of the tool do not discriminate against any particular group or community. AI tools can become biased due to the data used to train them, the algorithms used to process the data, or the way the tools are designed to be used.

To address these issues, it is important to use diverse and representative datasets to train AI tools. Data cleaning and preprocessing should be done to remove any bias that may be present in the data. It is also important to ensure that the algorithms used to process the data do not introduce any bias.

To ensure that AI tools are fair and unbiased, they should be tested using diverse and representative datasets to identify any biases that may be present. The results of the tool should be evaluated for fairness and any biases that are identified should be corrected.

It is also important to involve people from diverse backgrounds in the design and development of AI tools to ensure that the tools are designed to be inclusive and equitable. This can help to prevent bias from being unintentionally introduced into the tool during the design and development process.

Overall, it is important to approach the development and use of AI tools with a commitment to fairness and equity to ensure that these tools are used to benefit everyone.

Chapter 10:
Looking Ahead: ChatGPT and the Future of Programming

As a language model, I can say that the future of programming with AI and ChatGPT is exciting and full of possibilities. ChatGPT and other AI tools have already revolutionized the way we write, debug, and optimize code. In the future, we can expect even more advanced AI systems that can analyze code more deeply, identify patterns, and suggest more complex and optimized solutions. AI systems can also help make programming more accessible to non-experts, enabling more people to learn and use programming languages in their daily lives. Additionally, AI can help address issues of code security and privacy, reducing the risks of hacking and data breaches. However, it is important to continue to address ethical concerns and ensure responsible use of AI in programming to avoid negative consequences.

Current Trends and Future Predictions

Current trends suggest that AI-assisted programming is becoming more prevalent in the industry, with many tools and platforms being developed to assist developers in writing, testing, and maintaining code. As AI technology continues to advance, it is likely that these tools will become even more sophisticated and capable of handling more complex tasks.

In the future, it is possible that AI systems like ChatGPT may be used to generate entire software applications, automating much of the development process. However, it is also important to consider the ethical implications of such a scenario, including the potential impact on jobs in the software development industry.

Overall, the future of programming with AI is likely to involve a mix of human and machine intelligence working together to create more efficient, reliable, and effective software. As such, it is important for programmers to embrace these new technologies and explore ways to incorporate them into their work in responsible and ethical ways.

Preparing for an AI-Driven Programming Landscape

As the use of AI continues to grow in programming, it is important for programmers to stay up-to-date with the latest developments in the field. This includes learning about new AI tools and techniques, as well as staying informed about ethical considerations and responsible use of AI in programming.

One way to prepare for an AI-driven programming landscape is to engage in ongoing professional development and training. This may involve attending conferences and workshops, participating in online courses and communities, and collaborating with other programmers and experts in the field.

Another important consideration is to continue developing strong foundational programming skills, such as algorithm design, data structures, and software engineering principles. While AI tools can assist with many aspects of programming, they are not a substitute for a strong understanding of these foundational concepts.

Finally, programmers should also be aware of the potential impact of AI on the job market and the future of work. While AI may automate certain tasks and processes, there will likely always be a need for skilled programmers who can design, implement, and maintain AI systems. By staying informed and adaptable, programmers can position themselves for success in an AI-driven future.

The best PROMPTs for the PROGRAMMER:

Write in GPT Chat this prompt exactly as it is written below. Then try changing the terms you find in the " " to get the work that works best for you. Remember that in case you are a programmer, this prompt is PERFECT for you, it is a very powerful tool make good use of it! Prompt:

You are a programmer and will act like one. You will write a code in "Python" for create one app about "environmental pollution". The app will have the function of "illustrating Environmental Pollution in Europe for each country".

You are a programmer and will act like one. You will write a code in "HTML" for create one static website about "environmental pollution". The website will have the function of "illustrating Environmental Pollution in Europe for each country".

You are a seasoned programmer with over 20 years of experience and will act like one. Debug this code: "Copy and paste the code here and delete the quotes."

BOOK 7: "ChatGPT for Professionals: Efficiency, Time Management and Organization": Explores how ChatGPT can be used to improve professional productivity.

Chapter 1:
Introduction to ChatGPT for Professionals

ChatGPT is a powerful language model developed by OpenAI that can process and generate human-like text. It is based on the GPT-3.5 architecture and has been trained on a massive corpus of text from the internet. ChatGPT can understand natural language and generate responses that are contextually appropriate and grammatically correct. This technology has a wide range of potential applications, including market research, content creation, customer engagement, and more.

Professionals from various fields can use ChatGPT to enhance their work and streamline their processes. In this context, ChatGPT can be used as a tool to automate repetitive tasks, generate new ideas, and provide expert insights. For example, businesses can use ChatGPT to analyze market data, generate new product ideas, and create compelling marketing content. Researchers can use ChatGPT to assist in literature reviews, hypothesis generation and validation, and data analysis. Educators can use ChatGPT to create personalized learning experiences for students and improve communication with parents.

However, the use of AI technology like ChatGPT also raises ethical concerns and requires responsible use. Professionals must consider issues like bias and fairness, privacy and security, and accountability when using AI in their work. As AI technology continues to advance, it is important for professionals to stay up-to-date on current trends and best practices to ensure they are using these tools effectively and responsibly.

Understanding ChatGPT

ChatGPT is an artificial intelligence language model developed by OpenAI. It is designed to generate human-like text in response to natural language prompts. ChatGPT is based on deep learning techniques and is pre-trained on vast amounts of text data, allowing it to generate high-quality responses to a wide range of prompts. It can be used for various applications, including natural language processing, content generation, language translation, and conversational interfaces. As a language model, ChatGPT is able to understand the meaning and context of natural language, and it can generate responses that are coherent and relevant to the given prompt.

The Role of AI in Professional Productivity

AI can have a significant impact on professional productivity. By automating routine tasks, analyzing data, and providing insights and recommendations, AI tools like ChatGPT can help professionals work more efficiently and make better decisions. AI can also enhance collaboration and communication, allowing teams to work together more effectively regardless of their physical location. Additionally, AI can assist in complex problem-solving and decision-making, enabling professionals to address complex challenges and capitalize on emerging opportunities. Overall, AI has the potential to transform the way professionals work and help them achieve better results.

Chapter 2:
Efficiency and Automation with ChatGPT

ChatGPT can help improve professional productivity by providing efficient and automated solutions to various tasks. It can automate routine and repetitive tasks, allowing professionals to focus on more complex and high-value work. It can also help streamline workflows and improve efficiency by suggesting process improvements and identifying bottlenecks. Additionally, it can help professionals access information more quickly and easily, reducing the time required for research and decision-making. Overall, ChatGPT can help professionals achieve more in less time and with less effort, enabling them to be more productive and effective in their work.

Automating Routine Tasks

Automating routine tasks can save time, reduce errors, and increase efficiency in the workplace. ChatGPT can be used to automate routine tasks by training it to recognize patterns and respond to certain inputs with predefined actions. For example, you could use ChatGPT to automate tasks such as email filtering, data entry, appointment scheduling, and more. By automating these tasks, you can free up time to focus on more important tasks that require human intervention. Additionally, automation can improve accuracy and reduce errors, leading to improved productivity and job satisfaction.

Streamlining Workflows with AI

Streamlining workflows with AI involves automating repetitive and time-consuming tasks to improve productivity and efficiency. AI-powered tools such as ChatGPT can assist professionals by performing tasks like scheduling appointments, managing emails, generating reports, and summarizing documents. By freeing up time and resources, professionals can focus on higher-value tasks, leading to increased productivity and better decision-making. Additionally, AI can help identify areas of inefficiency and suggest improvements, allowing for continuous process improvement.

Chapter 3:
Time Management with ChatGPT

Time management is an important aspect of productivity, and ChatGPT can be a useful tool to help professionals manage their time more effectively. With its natural language processing capabilities, ChatGPT can help professionals automate routine tasks, prioritize work, and schedule appointments and meetings. Additionally, ChatGPT can provide suggestions for optimizing workflow and improving productivity, based on the user's goals and work patterns. By using ChatGPT to manage their time more efficiently, professionals can free up more time for creative and strategic work, leading to greater overall productivity and success.

Prioritizing Tasks with AI Assistance

ChatGPT can assist professionals in prioritizing their tasks by analyzing their to-do lists and providing recommendations based on factors such as urgency, importance, and deadline. The AI can also suggest optimal scheduling of tasks based on the time required to complete them and the available time slots in the professional's calendar.

Additionally, ChatGPT can help professionals better manage their time by providing reminders and alerts for upcoming deadlines and meetings, as well as by automatically scheduling recurring tasks, such as regular reports or meetings.

By automating these time-consuming and repetitive tasks, professionals can focus their energy and attention on higher-value activities that require their expertise and creativity, ultimately improving their productivity and performance.

Scheduling and Calendar Management

Scheduling and calendar management are important aspects of professional productivity, and ChatGPT can be a valuable tool in this regard. Here are some ways ChatGPT can assist with scheduling and calendar management:

1. Scheduling meetings: ChatGPT can help schedule meetings by reviewing calendars and finding mutually available times. It can also send meeting invites and reminders to attendees.
2. Time blocking: ChatGPT can help users plan out their day by creating a schedule and blocking off time for specific tasks. This can help ensure that important tasks are completed and that time is used efficiently.

3. Calendar management: ChatGPT can help manage calendars by adding and updating events, setting reminders, and sending notifications.
4. Rescheduling appointments: ChatGPT can help users reschedule appointments by reviewing calendars and finding new available times.
5. Time zone conversion: ChatGPT can assist with time zone conversions when scheduling meetings or events with participants in different time zones.
6. Personalized scheduling preferences: ChatGPT can learn users' preferences and use this information to suggest scheduling options that align with their work style and preferences.

Overall, ChatGPT can assist with scheduling and calendar management by automating routine tasks, providing reminders, and suggesting scheduling options that align with users' preferences and work style. This can help users save time and be more productive in their professional lives.

Chapter 4:
Organizing Information with ChatGPT

ChatGPT can be a useful tool for organizing information in a professional context. It can be used to categorize and label documents, files, and other types of information, making it easier to search for and retrieve specific items. It can also be used to create summaries and abstracts of longer documents, saving time for professionals who need to review large amounts of information quickly.

Additionally, ChatGPT can help professionals stay on top of the latest news and trends in their industry by monitoring social media, news outlets, and other sources of information. It can filter and sort relevant information and provide alerts when new information becomes available.

Overall, ChatGPT can help professionals stay organized and informed, allowing them to focus on their core responsibilities and be more productive.

Using AI for Data Organization and Retrieval

AI can be very useful for organizing and retrieving data efficiently. ChatGPT can be used to automatically categorize data based on its content, making it easier to search and retrieve later on. This can save time and increase productivity in many industries, including healthcare, finance, and legal services.

In addition, ChatGPT can be used to create customized knowledge management systems that help professionals store and access information more effectively. These systems can be used to automate processes such as data entry, information retrieval, and knowledge sharing, allowing professionals to focus on more complex tasks.

Overall, using ChatGPT for data organization and retrieval can help professionals save time, reduce errors, and improve the accuracy and consistency of their work.

Note-Taking and Meeting Summaries with ChatGPT

ChatGPT can assist in taking notes and summarizing meetings. During meetings, ChatGPT can transcribe the conversation in real-time and convert it into text format. It can also identify the key points discussed in the meeting and summarize them for quick reference later. This can save time and effort for busy professionals who need to review the meeting notes at a later time. Additionally, ChatGPT can help in generating meeting minutes and action items, which can be shared with the participants for review and follow-up.

Chapter 5:
Communication and Collaboration

Communication and collaboration are critical aspects of productivity in any workplace. ChatGPT can be a useful tool for enhancing communication and collaboration among professionals.

One way ChatGPT can help is by generating summaries of meetings or discussions. This can save time for busy professionals who may not have the time to listen to entire recordings or read through lengthy transcripts. With ChatGPT, they can quickly generate summaries that capture the key points of the discussion.

ChatGPT can also help improve communication by providing quick responses to common questions or concerns. For example, an HR department might use ChatGPT to generate responses to common employee questions about benefits, policies, or procedures. This can help reduce the workload on HR staff while still ensuring that employees receive the information they need.

Another way ChatGPT can improve collaboration is by facilitating knowledge sharing among team members. ChatGPT can be used to generate summaries or insights from large amounts of data or research, which can be shared with other team members. This can help ensure that everyone on the team has access to the same information and can make informed decisions.

Finally, ChatGPT can be used to facilitate communication and collaboration among remote teams. With ChatGPT, team members can communicate with each other in real-time, share information and data, and collaborate on projects regardless of their physical location. This can help improve productivity and ensure that everyone on the team is working toward the same goals.

Enhancing Team Communication with ChatGPT

ChatGPT can be used to enhance team communication in various ways. For instance, it can be used to facilitate real-time translation of conversations among team members who speak different languages. It can also be used to transcribe audio or video recordings of meetings and convert them into text, which can then be easily shared and reviewed by team members.

Moreover, ChatGPT can be used to create chatbots that can be integrated with team communication platforms, such as Slack or Microsoft Teams, to provide instant answers to frequently asked questions or provide status updates on ongoing projects. This can reduce the time spent on repetitive tasks and increase overall productivity.

Additionally, ChatGPT can be used to generate summaries of long emails, reports, or articles, which can then be shared with team members who need to be informed of the contents without having to go through the entire document. This can save time and ensure that everyone is on the same page.

Lastly, ChatGPT can be used to analyze the tone and sentiment of team communications, such as emails or chat messages, to identify potential conflicts or areas where misunderstandings may arise. This can help team members address issues proactively and work collaboratively to prevent or resolve conflicts.

Using AI for Collaborative Projects and Brainstorming

ChatGPT can be a valuable tool for collaborative projects and brainstorming. One way to use ChatGPT for this purpose is to have team members input ideas or questions, and then use ChatGPT to generate responses or additional ideas. This can be useful for generating new ideas, solving problems, and exploring different approaches to a project.

Additionally, ChatGPT can assist in facilitating discussions and debates. For example, team members can pose questions or topics to ChatGPT, which can generate responses that can then be used as a starting point for further discussion. This can help to promote constructive dialogue and ensure that all team members have an opportunity to contribute to the conversation.

Overall, ChatGPT can be a useful tool for promoting collaboration and communication within a team, and can help to generate new ideas and approaches to projects.

Chapter 6:
Professional Learning and Development

Professional learning and development refer to the continuous improvement of an individual's knowledge, skills, and abilities related to their profession or occupation. ChatGPT can be a useful tool to aid professionals in their learning journey. Here are some ways in which ChatGPT can be utilized for professional learning and development:

1. Research: ChatGPT can help professionals in their research by assisting in literature review, hypothesis generation, and data analysis.
2. Knowledge acquisition: ChatGPT can provide professionals with information on a wide range of topics and help them learn new concepts and terminologies related to their field.
3. Problem-solving: ChatGPT can be used to help professionals solve complex problems by providing suggestions and recommendations based on data and research.
4. Skill development: ChatGPT can help professionals develop new skills by providing personalized learning experiences and recommending resources.
5. Training and education: ChatGPT can be used as a tool for training and education, providing professionals with interactive learning experiences and personalized feedback.
6. Networking and collaboration: ChatGPT can connect professionals with others in their field and facilitate collaboration, knowledge sharing, and peer-to-peer learning.
7. Personal growth and development: ChatGPT can be used to support professionals in their personal growth and development by providing personalized recommendations for books, articles, and other resources.

Overall, ChatGPT can be a valuable tool for professionals looking to enhance their knowledge, skills, and abilities, as well as their overall effectiveness in their profession or occupation.

Utilizing ChatGPT for Skill Development

ChatGPT can be an effective tool for professionals looking to develop their skills. Here are some ways in which ChatGPT can be utilized for professional learning and development:

1. Knowledge acquisition: ChatGPT can be used to learn about a wide range of topics, from industry-specific concepts to general business principles. By asking questions and having conversations with ChatGPT, professionals can gain a deeper understanding of complex subjects

and stay up-to-date on the latest trends in their field.

2. Professional writing: ChatGPT can help professionals improve their writing skills by providing feedback and suggesting revisions. Whether it's a report, a proposal, or an email, ChatGPT can offer insights on grammar, style, and tone to ensure that written communication is clear, concise, and effective.

3. Presentation skills: ChatGPT can assist professionals in preparing for presentations by offering suggestions on content, organization, and delivery. Professionals can practice their presentations with ChatGPT, receive feedback on their performance, and make adjustments accordingly.

4. Technical skills: ChatGPT can be used to learn technical skills such as coding, data analysis, and graphic design. By asking questions and receiving feedback from ChatGPT, professionals can hone their skills and stay competitive in their industry.

5. Leadership development: ChatGPT can be used to develop leadership skills by providing insights on managing people, making decisions, and communicating effectively. Professionals can ask ChatGPT for advice on specific situations or for general guidance on becoming a more effective leader.

Overall, ChatGPT can be a valuable tool for professional learning and development. By leveraging the power of AI, professionals can enhance their skills, acquire new knowledge, and become more effective in their roles.

Personalized Learning Plans with AI

Personalized learning plans with AI refer to using artificial intelligence to help professionals create customized plans for their own learning and development. By analyzing data from their work and learning history, AI can help professionals identify their strengths and weaknesses, and then suggest learning opportunities that align with their interests and goals. With this approach, professionals can focus their learning efforts on areas where they need the most improvement, and enhance their skills in a way that is personalized and efficient.

ChatGPT can be used to create personalized learning plans by providing insights into the specific skills and knowledge areas that a professional needs to develop. For example, ChatGPT can analyze a professional's work history and recommend learning opportunities based on the specific tasks they have performed, the tools and technologies they have used, and the outcomes they have achieved. By doing so, ChatGPT can help professionals build on their existing knowledge and experience, and improve their skills in a targeted way.

Additionally, ChatGPT can provide suggestions for learning resources based on the professional's learning style and preferences. For example, some professionals may prefer video tutorials or interactive quizzes, while others may prefer written articles or hands-on exercises. By analyzing a professional's learning history and preferences, ChatGPT can recommend resources that are more likely to be engaging and effective.

Overall, using ChatGPT for personalized learning plans can help professionals optimize their learning and development efforts, and enhance their skills in a way that is tailored to their individual needs and preferences.

Chapter 7:
Work-Life Balance with ChatGPT

ChatGPT can also assist in achieving work-life balance by automating routine tasks, organizing information and providing reminders. It can help professionals prioritize their tasks, manage their time effectively and reduce stress. By using AI to automate mundane and repetitive tasks, professionals can focus on more important and meaningful work. Additionally, AI tools can help professionals achieve a better work-life balance by providing suggestions for breaks, rest periods and leisure activities based on their schedule and preferences. This can help prevent burnout and increase job satisfaction.

Using AI to Manage Personal and Professional Life

AI can be useful in managing both personal and professional life. ChatGPT can help individuals to prioritize their tasks and manage their time effectively. This can enable them to focus on important tasks and delegate less important ones. By automating routine tasks, individuals can save time and effort, which can be used for other purposes. Additionally, ChatGPT can assist individuals in setting personal and professional goals and create a plan to achieve them. This can help individuals to stay motivated and focused. Finally, ChatGPT can also help individuals to develop new skills and improve their knowledge through personalized learning plans. By utilizing AI tools like ChatGPT, professionals can maintain a work-life balance while achieving their personal and professional goals.

Mindfulness and Stress Management with ChatGPT

ChatGPT can provide assistance to professionals in managing their stress levels and practicing mindfulness. It can suggest mindfulness exercises, provide guided meditations, and offer tips for stress reduction. Additionally, ChatGPT can help professionals manage their workloads more effectively, allowing them to feel more in control and reduce feelings of overwhelm. It can also provide reminders for breaks and encourage professionals to take time for self-care activities, such as exercise or spending time with loved ones. Overall, ChatGPT can be a helpful tool for professionals looking to prioritize their mental health and wellbeing.

Chapter 8:
ChatGPT for Remote Work

ChatGPT can be an incredibly useful tool for remote work. As more and more companies shift to remote work setups, the need for digital collaboration and communication tools has increased. ChatGPT can assist in a number of ways:

1. Collaboration: ChatGPT can be used to facilitate virtual meetings, brainstorming sessions, and project collaborations. ChatGPT can also help teams work together on a shared document, such as a project plan or proposal.
2. Time Management: ChatGPT can help remote workers keep track of their time by scheduling and prioritizing tasks. It can also remind them of upcoming deadlines or meetings.
3. Personalized Learning: ChatGPT can assist in personalized professional development, providing recommendations for courses or books to read based on the employee's interests and career goals.
4. Wellness and Work-Life Balance: ChatGPT can be used to schedule wellness breaks and remind employees to take breaks or step away from work when needed.
5. Virtual Assistance: ChatGPT can act as a virtual assistant, answering questions or providing reminders for important tasks.

Overall, ChatGPT can enhance productivity and support remote workers in a number of ways, making it a valuable tool for any team working remotely.

Enhancing Productivity in Remote Settings

With the growing popularity of remote work, professionals are seeking new ways to increase their productivity outside of the traditional office setting. ChatGPT can be a valuable tool in this regard by providing a range of benefits for remote workers.

One advantage of using ChatGPT for remote work is the ability to automate routine tasks, such as scheduling and data entry. By delegating these tasks to ChatGPT, professionals can save time and focus on more high-level work.

Another benefit of ChatGPT for remote work is improved collaboration and communication. ChatGPT can facilitate discussions, brainstorming sessions, and even meetings, helping remote teams stay connected and on track.

ChatGPT can also assist with time management, prioritizing tasks, and organizing information. These features can help remote workers stay on top of their workload and manage their time effectively.

Finally, ChatGPT can be used for personalized learning and development, allowing remote workers to develop new skills and stay up to date on the latest industry trends.

Overall, ChatGPT can be an invaluable tool for remote workers looking to enhance their productivity and stay connected with their team.

Adapting to Virtual Collaboration and Communication

As remote work becomes more common, professionals must adapt to new modes of collaboration and communication. ChatGPT can play a key role in enhancing productivity in remote settings by facilitating virtual communication and collaboration. Here are some ways in which ChatGPT can help professionals working remotely:

1. Virtual meetings: ChatGPT can be used to facilitate virtual meetings, allowing team members to communicate and collaborate in real-time. The AI can be used to create meeting agendas, schedule meetings, and even take notes during the meeting.
2. Collaborative work: ChatGPT can be used to facilitate collaborative work on projects, allowing team members to work together on documents and other materials in real-time.
3. Communication: ChatGPT can be used to facilitate communication between team members, allowing for quick and easy exchange of information.
4. Time management: ChatGPT can be used to manage time and keep team members on track with deadlines and schedules.
5. Remote training: ChatGPT can be used to provide remote training to team members, allowing them to learn new skills and improve their productivity.
6. Personalized support: ChatGPT can provide personalized support to remote workers, helping them to stay motivated and on track with their work.

Overall, ChatGPT can play a critical role in enhancing productivity and communication in remote work settings, allowing teams to work together effectively and efficiently, regardless of their physical location.

Chapter 9:
Ethical Considerations in AI-Assisted Work

When it comes to AI-assisted work, ethical considerations are critical to ensuring that the technology is used responsibly and fairly. Here are some of the key ethical considerations that professionals should keep in mind when using ChatGPT or any other AI tool:

1. Transparency: It's important to be transparent about how AI is being used and what data is being collected. This includes providing clear explanations of how AI algorithms work and how they are being used to make decisions.
2. Bias: AI tools can perpetuate bias if they are trained on biased data or if they are not designed to account for diversity. It's important to ensure that AI algorithms are fair and unbiased.
3. Privacy: AI tools often rely on data, which can include personal and sensitive information. It's essential to ensure that data is collected, stored, and used responsibly and in accordance with relevant privacy laws and regulations.
4. Responsibility: As with any tool, it's essential to use AI responsibly and with a clear understanding of its capabilities and limitations. Professionals should be aware of the potential consequences of using AI tools and take responsibility for any unintended consequences.
5. Human oversight: AI tools should be used as an aid, not a replacement, for human decision-making. Professionals should exercise human judgment and oversight when using AI tools and ensure that they are being used in a way that aligns with organizational values and goals.

By keeping these ethical considerations in mind, professionals can use AI tools like ChatGPT in a way that benefits both themselves and society as a whole.

Responsible Use of AI in the Workplace

As with any technology, the use of AI in the workplace comes with ethical considerations that must be addressed to ensure responsible use. Here are some key considerations:

1. Data privacy: AI tools rely on data to function, and this data must be collected from somewhere. It's important to ensure that data collection is done in a responsible and transparent way, and that individuals' privacy rights are respected.
2. Bias: AI algorithms are only as good as the data they're trained on, and if that data is biased, the algorithm will be too. It's important to be aware of potential biases in the data and work to

correct them.

3. Transparency: AI algorithms can be difficult to interpret, which can make it challenging to understand why a particular decision was made. It's important to ensure that AI decisions are transparent and explainable, so that individuals can understand why a decision was made.
4. Job displacement: AI has the potential to automate many tasks that are currently done by humans, which could lead to job displacement. It's important to consider the impact of AI on employment and work to create policies that mitigate negative effects.
5. Accountability: As AI is increasingly used to make decisions that have significant impact on individuals, it's important to ensure that there is accountability for those decisions. This can include creating standards for AI systems, and establishing processes for challenging and correcting AI decisions that are incorrect or unfair.

Overall, it's important to approach the use of AI in the workplace with care and consideration, and to work to ensure that it is used in a responsible and ethical way.

Addressing Equity and Access Issues

As organizations continue to adopt AI tools and technologies, it is important to ensure that equity and access issues are addressed. The benefits of AI should not be limited to certain groups of people or organizations. Here are some ways to address equity and access issues in AI-assisted work:

1. Ensure that AI tools are accessible to everyone: Organizations should make sure that AI tools are accessible to everyone regardless of their race, gender, age, or any other characteristic. This can be achieved by providing training and resources to employees and making sure that the tools are user-friendly.
2. Monitor and address biases: AI tools can inherit biases from the data that is used to train them. It is important to monitor and address any biases that are identified. This can be achieved by regularly reviewing the data that is used to train the tools and by having diverse teams of people involved in the development of AI tools.
3. Address the digital divide: There are still many people who do not have access to the internet or the necessary technology to use AI tools. Organizations can address this by providing access to technology and the internet to their employees or by partnering with community organizations to provide access to people in the community.
4. Ensure that AI is used ethically: Organizations should have policies in place to ensure that AI is used ethically. This includes ensuring that AI is not used to discriminate against people, that privacy is protected, and that the use of AI is transparent.

5. Involve diverse perspectives in decision-making: When making decisions about the use of AI, it is important to involve people with diverse perspectives. This can help to identify and address potential issues related to equity and access.

Chapter 10:
Looking Ahead: ChatGPT and the Future of Work

As AI technology continues to advance, it is likely that ChatGPT and similar language models will play an increasingly important role in the workplace. The ability of AI to automate routine tasks, streamline workflows, and provide personalized assistance could significantly enhance professional productivity and performance.

Additionally, as more companies move towards remote work and virtual collaboration, AI-powered tools such as ChatGPT could help to facilitate communication and collaboration in distributed teams.

However, as with any new technology, there are ethical considerations that must be taken into account. It will be important to ensure that the use of AI in the workplace is transparent, equitable, and respectful of individual privacy.

Overall, the future of work with ChatGPT and other AI tools is likely to be one of increased efficiency, productivity, and collaboration.

Current Trends and Future Predictions

Some of the current trends in AI and the workplace include the use of natural language processing for customer service chatbots, the automation of routine tasks through robotic process automation (RPA), and the use of machine learning for predictive analytics and data-driven decision-making. In the future, AI is expected to continue to transform the way we work, with advancements in areas such as personalized learning, virtual and augmented reality, and the use of AI-powered assistants to enhance productivity and efficiency.

As AI becomes more integrated into the workplace, there will likely be a shift in the skills and competencies that are valued. Soft skills such as creativity, emotional intelligence, and adaptability will become increasingly important, as they are less likely to be automated. Workers will also need to be comfortable with using and working alongside AI systems, and will need to develop the skills to interact with and manage these systems.

Overall, AI has the potential to improve productivity and efficiency in the workplace, but it is important to approach its use in a responsible and ethical manner, and to consider the potential impact on workers and society as a whole.

Preparing for an AI-Driven Work Landscape

The increasing use of AI in the workplace is changing the way we work, and this trend is expected to continue in the future. As AI technology advances, it has the potential to revolutionize many aspects of work, including automation, personalization, and decision-making.

Some current trends in AI in the workplace include:

1. Automation of routine tasks: AI-powered automation tools are increasingly being used to perform routine, repetitive tasks, such as data entry and processing.
2. Personalization of work experience: AI can help to tailor work experiences to individual needs and preferences, allowing employees to work more efficiently and effectively.
3. Decision-making support: AI can help to analyze large amounts of data and provide insights that can aid decision-making.
4. Remote work enablement: AI-powered communication and collaboration tools are helping to facilitate remote work.

In the future, AI is likely to play an even bigger role in the workplace. Some potential developments include:

1. Increased automation: As AI technology advances, it will become increasingly capable of automating more complex tasks.
2. Improved personalization: AI will continue to improve its ability to understand individual needs and preferences, leading to more personalized work experiences.
3. Enhanced decision-making support: AI will become even more capable of analyzing large amounts of data and providing insights that aid decision-making.
4. Greater integration with other technologies: AI will become more integrated with other technologies, such as the Internet of Things and blockchain, leading to more powerful and efficient systems.

As AI becomes more prevalent in the workplace, it is important to ensure that it is used responsibly and ethically, with attention paid to issues such as bias and privacy. Organizations will need to carefully consider how they can use AI to enhance productivity and improve the work experience while also minimizing potential negative impacts.

BOOK 8: "ChatGPT for Social Media Managers: Content Generation, Engagement and Analysis": Shows how ChatGPT can be used in the context of social media.

Chapter 1:
Introduction to ChatGPT for Social Media Managers

Social media has become an integral part of businesses' marketing strategies, and social media managers play a crucial role in the success of these campaigns. Managing social media platforms can be time-consuming and requires constant attention. That's where artificial intelligence (AI) comes in, and ChatGPT is a powerful tool that can assist social media managers in several ways.

ChatGPT is a large language model that uses deep learning techniques to generate human-like responses to natural language prompts. It has been trained on vast amounts of data and can recognize patterns in the text, which makes it a useful tool for social media managers who need to analyze large amounts of data quickly.

In this guide, we will explore how ChatGPT can be used to enhance social media management and improve social media campaigns' overall effectiveness. We will discuss various ways social media managers can leverage ChatGPT's capabilities, including:

1. Idea generation and content creation
2. Social listening and sentiment analysis
3. Personalized customer engagement
4. Advertising and targeting
5. Performance monitoring and reporting
6. Ethical considerations in AI-assisted social media management

By understanding how ChatGPT can help social media managers improve their workflows and campaigns, we can explore the full potential of AI in social media management.

Understanding ChatGPT

ChatGPT is an artificial intelligence language model developed by OpenAI that uses deep learning techniques to generate human-like language. It is capable of generating text in response to a given prompt or query and can perform a wide range of language-related tasks, including natural language processing, language translation, question-answering, and conversation generation. ChatGPT is trained on a massive corpus of text data from the internet, allowing it to generate highly relevant and coherent responses to a wide range of prompts. As such, it has a wide range of potential applications across various industries and domains, including social media management.

The Role of AI in Social Media Management

ChatGPT can play an important role in social media management by assisting with content creation, scheduling, and analysis. Social media managers can use ChatGPT to generate post ideas, captions, and hashtags. They can also use the tool to schedule posts at optimal times for engagement and monitor their social media performance metrics.

AI can also assist in analyzing audience sentiment, identifying trending topics and hashtags, and detecting fake news and misinformation. ChatGPT can help social media managers stay up-to-date with industry news and social media trends, as well as identify opportunities for influencer partnerships and collaborations.

Furthermore, ChatGPT can aid in developing chatbots and virtual assistants for social media platforms. Social media managers can use the tool to develop and train chatbots to provide customer service, answer frequently asked questions, and automate responses.

Overall, ChatGPT can help social media managers save time and increase their efficiency in managing their social media accounts, while also providing valuable insights and improving customer engagement.

Chapter 2:
Content Generation with ChatGPT

Social media is all about content creation, curation, and dissemination. ChatGPT can help social media managers in creating engaging and effective content for their social media channels. ChatGPT can suggest topics, write captions, and even generate complete social media posts. It can analyze social media data and identify trends and topics that are currently popular among the target audience. Additionally, ChatGPT can also provide inspiration for visual content, such as images and videos. This can help social media managers save time and effort in content creation, allowing them to focus on other important tasks.

Creating Engaging Posts with AI

As a social media manager, you can use ChatGPT to generate engaging posts for your social media channels. With its advanced language processing capabilities, ChatGPT can analyze your target audience's interests, engagement metrics, and preferred style of content to create posts that are tailored to their preferences.

To use ChatGPT for content generation, you can provide it with some basic information about your audience, such as age, gender, location, and interests. You can also provide some general topics or themes that you want the content to cover.

Based on this information, ChatGPT can generate a list of potential post ideas, headlines, and even full-length posts. You can then review and edit these suggestions to ensure they align with your brand's tone and messaging.

Using ChatGPT for content generation can save you time and resources while also improving the effectiveness of your social media efforts.

Writing Captions, Tweets, and Other Social Content

ChatGPT can be a helpful tool for social media managers in generating captions, tweets, and other social content. By providing prompts or keywords, ChatGPT can suggest ideas for social media posts that align with a brand's voice and messaging. This can save time and effort in the content creation process, while also generating new and innovative ideas.

Social media managers can also use ChatGPT to experiment with different writing styles and tones. For example, if a brand is known for using a humorous tone on social media, ChatGPT can suggest funny phrases or puns to incorporate into posts.

However, it's important to note that while ChatGPT can be a useful tool for generating ideas, it's ultimately up to the social media manager to review and edit the content to ensure it aligns with the brand's values and messaging.

Chapter 3:
Audience Engagement with ChatGPT

ChatGPT can play an important role in audience engagement for social media managers. By analyzing social media data, ChatGPT can generate insights about audience preferences, interests, and behavior. This information can be used to create more targeted and personalized content that resonates with the audience.

ChatGPT can also assist with social media listening, which involves monitoring social media channels for mentions of a brand, product, or service. By analyzing these mentions, ChatGPT can identify trends and sentiment, which can help social media managers to better understand their audience and tailor their content accordingly.

Additionally, ChatGPT can be used to generate responses to customer inquiries and comments on social media platforms. By analyzing the content of the inquiry or comment, ChatGPT can generate a response that is tailored to the specific situation and context. This can help social media managers to respond more quickly and efficiently to customer inquiries, which can lead to increased customer satisfaction and loyalty.

Automating Response to Comments and Messages

Automating response to comments and messages can be a helpful way to manage the volume of incoming messages on social media platforms. ChatGPT can assist with this task by generating personalized responses based on the content of the incoming messages. The AI model can be trained on existing messages and responses to learn how to generate appropriate responses for different types of inquiries or feedback.

ChatGPT can also help to identify messages that require urgent attention or escalation to a human representative. By analyzing the content of the message, the AI model can determine the level of urgency and direct it to the appropriate person or department for handling.

Overall, using ChatGPT to automate response to comments and messages can help social media managers save time and ensure that all messages are addressed in a timely manner.

Encouraging Interaction with AI

One way to encourage interaction with AI is by setting up automated messages or chatbots that respond to users in a conversational manner. These automated systems can help answer common questions or provide basic information to users, freeing up time for social media managers to focus on more complex tasks. Additionally, AI tools can be used to analyze user data and preferences to better target content and improve engagement. For example, machine learning algorithms can identify patterns in user behavior to suggest the best times to post or the types of content that are most likely to resonate with specific segments of the audience.

Chapter 4:
Social Media Analytics with ChatGPT

Social media analytics is an important part of social media management, and AI can help streamline the process. ChatGPT can assist social media managers by analyzing data to provide insights into user behavior, sentiment analysis, and engagement metrics. This information can be used to optimize social media content and inform social media strategies. ChatGPT can also help with the creation of social media reports and dashboards, providing a snapshot of social media performance and allowing for real-time monitoring and adjustment of social media campaigns. Overall, AI tools like ChatGPT can greatly improve social media management and help businesses make data-driven decisions.

Analyzing Engagement Metrics with AI

Social media managers can use AI-powered tools such as ChatGPT to analyze engagement metrics and improve their social media strategies. These tools can help analyze social media data and provide insights on what content is performing well and how to optimize future posts.

ChatGPT can be trained on past engagement data to learn what factors contribute to higher engagement rates. For example, it can learn which types of posts are more likely to receive likes, comments, and shares. Using this knowledge, ChatGPT can generate recommendations for optimizing social media content, such as using certain keywords or posting at specific times of day.

Additionally, ChatGPT can help social media managers create more accurate and useful reports on engagement metrics. By analyzing data and providing insights, ChatGPT can help social media managers make informed decisions and improve their social media strategies.

Understanding Audience Sentiment with ChatGPT

ChatGPT can be used for sentiment analysis on social media data. Sentiment analysis is the process of determining whether a piece of text expresses a positive, negative, or neutral sentiment. With the help of machine learning algorithms, ChatGPT can analyze large amounts of social media data to determine the sentiment of a particular topic or brand. This can be useful for social media managers to track the sentiment of their brand, identify potential issues or areas for improvement, and adjust their strategy accordingly. By analyzing the sentiment of social media data, ChatGPT can help social media managers make data-driven decisions that are more likely to resonate with their target audience.

Chapter 5:
Social Media Strategy and Planning

ChatGPT can be a helpful tool for social media managers in creating and planning social media strategies. Here are some ways ChatGPT can be used in this area:

1. Generating Ideas for Social Media Posts: ChatGPT can be used to generate ideas for social media posts. It can provide inspiration for content creation and help social media managers come up with unique and creative ideas.
2. Developing a Content Calendar: ChatGPT can help in creating a content calendar for social media platforms. By providing suggestions for topics to cover, ChatGPT can help in developing a consistent schedule for posting on social media.
3. Trend Analysis: ChatGPT can analyze trends in social media and help social media managers keep track of popular topics and hashtags. This can help in developing content that is relevant to the target audience.
4. Personalization: ChatGPT can analyze user data and preferences to help personalize social media content. This can help in creating content that resonates with the audience and encourages engagement.
5. Targeting Specific Demographics: ChatGPT can help in targeting specific demographics on social media. By analyzing user data, it can provide insights into the target audience's interests and preferences, which can be used to create content that is tailored to their needs.

Overall, ChatGPT can be a useful tool for social media managers in developing and implementing social media strategies. By providing insights and suggestions, ChatGPT can help in creating engaging and effective social media content.

Developing a Social Media Plan with ChatGPT

Developing a social media plan can be a daunting task, but ChatGPT can assist social media managers in creating a comprehensive strategy. Here are some ways that ChatGPT can be utilized:

1. Brainstorming ideas: ChatGPT can generate a list of potential content ideas based on a given topic or theme. By inputting keywords or phrases related to your brand or industry, ChatGPT can provide a list of content ideas that can be used to create a social media plan.
2. Identifying trends: Social media is constantly evolving, and it is important to stay on top of the

latest trends and topics. ChatGPT can analyze data from social media platforms to identify trending topics and hashtags that can be incorporated into a social media plan.

3. Defining target audience: ChatGPT can help social media managers define their target audience by analyzing demographic data and social media behavior. This information can be used to create content that is tailored to the interests and preferences of the target audience.
4. Creating a content calendar: ChatGPT can assist in creating a content calendar by generating ideas for specific days or events. For example, ChatGPT can suggest content ideas for holidays, product launches, or other important dates.
5. Monitoring and analyzing performance: ChatGPT can analyze social media data to track engagement and performance. This data can be used to adjust the social media plan and optimize content for maximum engagement.

Overall, ChatGPT can be a valuable tool for social media managers to create a comprehensive and effective social media plan.

Scheduling and Timing Posts with AI

ChatGPT can be used to schedule and time posts on social media platforms. With its natural language processing capabilities, ChatGPT can help social media managers identify the best times to post based on factors such as audience demographics, platform algorithms, and historical engagement data. Additionally, ChatGPT can be used to automate the scheduling of posts, allowing social media managers to focus on other aspects of their job. By using AI to optimize the timing and frequency of posts, social media managers can increase engagement and reach a larger audience.

Chapter 6:
Crisis Management and Damage Control

Social media can be a double-edged sword. While it is an excellent platform for businesses to reach a vast audience, one negative comment or review can spread like wildfire and severely impact a company's reputation. This is where crisis management and damage control come into play.

ChatGPT can assist social media managers in handling crises by:

1. Identifying potential crises: By monitoring social media activity, ChatGPT can flag posts or comments that could potentially spiral into a crisis.
2. Generating responses: Social media managers can use ChatGPT to generate response templates and customize them based on the specific crisis.
3. Identifying influencers: ChatGPT can help social media managers identify influential users who can help spread positive messaging during a crisis.
4. Monitoring sentiment: ChatGPT can analyze social media sentiment to gauge how users are reacting to the crisis and adjust messaging accordingly.
5. Responding quickly: ChatGPT can help social media managers respond quickly to a crisis by automating responses and providing real-time alerts.

Overall, ChatGPT can help social media managers stay ahead of crises and effectively manage damage control to protect a company's reputation.

Using ChatGPT to Monitor Social Media Chatter

ChatGPT can be used to monitor social media chatter by analyzing online conversations and identifying any negative or controversial posts related to a brand or organization. This can help social media managers to quickly identify potential issues and respond appropriately, either by addressing the concerns or taking other necessary actions to mitigate the situation.

Using natural language processing (NLP) techniques, ChatGPT can identify patterns and trends in social media conversations, detect sentiment, and analyze the tone and context of the discussion. This allows social media managers to quickly respond to any negative comments or complaints, as well as identify potential issues before they escalate.

In addition, ChatGPT can be used to generate responses to common questions and concerns, providing a faster and more consistent approach to handling customer inquiries. This can help improve customer satisfaction and reduce the workload of social media managers, allowing them to focus on more strategic tasks.

Responding to Negative Feedback with AI

AI can help to respond to negative feedback in different ways:

1. Automated response: AI can be trained to respond automatically to certain types of negative feedback. For example, if a customer complains about a delay in service, the AI system can be programmed to apologize and provide an estimated time for the service to be completed. This can help to address the customer's concerns quickly and efficiently.
2. Sentiment analysis: AI can be used to analyze social media posts and comments to determine the sentiment behind them. This can help social media managers to identify negative feedback and respond appropriately. For example, if a post is highly critical of a product, the social media manager can respond by acknowledging the customer's concerns and offering a solution.
3. Reputation management: AI can be used to monitor social media and other online platforms for negative feedback and potential reputation damage. The system can alert the social media manager when negative comments or reviews are posted, allowing them to respond quickly and minimize the impact of the feedback.
4. Personalized response: AI can help to personalize responses to negative feedback by analyzing the customer's previous interactions with the brand. This can help the social media manager to craft a response that is tailored to the customer's specific concerns and preferences.

In all cases, it is important to ensure that the AI system is trained and programmed to respond appropriately to negative feedback, taking into account the context and the specific needs of the customer. Additionally, it is important to have a human oversight to ensure that the AI system is performing as expected and that responses are consistent with the brand's values and messaging.

Chapter 7:
Brand Voice and Consistency

Brand Voice and Consistency is a vital aspect of social media management. It involves establishing a unique personality and tone for your brand and maintaining it consistently across all social media platforms. This helps build brand recognition, trust, and loyalty among your audience.

ChatGPT can assist social media managers in creating and maintaining a consistent brand voice by suggesting tone, language, and style that align with your brand's personality. It can analyze your existing social media content and suggest improvements to maintain a consistent voice.

Moreover, ChatGPT can also help in identifying inconsistencies in the brand voice, providing feedback on brand messaging, and proposing solutions to ensure that the voice is maintained across all channels. It can suggest modifications to existing content to make it more brand-aligned and recommend new content that complements the existing tone and style. This can significantly enhance brand recognition and reputation in the online world.

Maintaining a Consistent Brand Voice with ChatGPT

ChatGPT can be used to help maintain a consistent brand voice across social media platforms. By inputting sample text and messaging guidelines, the AI model can provide suggestions and guidance on how to write content that aligns with the brand's tone and style. It can also help identify and flag any deviations from the brand's voice, enabling social media managers to make necessary adjustments. Additionally, ChatGPT can assist in crafting messaging for new campaigns or product launches, ensuring that the brand's voice is maintained even as content needs to evolve to suit new goals and objectives.

Ensuring Content Consistency Across Platforms

ChatGPT can be used to ensure content consistency across different social media platforms. Social media managers can use it to generate content that is consistent with their brand voice, values, and guidelines across all their social media channels. By training the model with examples of brand-approved content, it can learn to recognize and replicate the style, tone, and messaging that the brand wants to convey.

With ChatGPT, social media managers can also ensure that their content is consistent with the platform-specific guidelines for each social media platform they are using. For example, they can train the model to generate tweets that conform to Twitter's character limit, or posts that meet Instagram's visual and caption requirements.

Moreover, ChatGPT can help ensure consistency in the timing and frequency of social media posts. By analyzing engagement data and identifying the optimal time for posting on each platform, the model can generate a schedule for when to publish content, ensuring that the content is seen by the maximum number of followers.

Chapter 8:
ChatGPT for Influencer Marketing

Influencer marketing is an essential part of social media marketing, and ChatGPT can be a valuable tool for influencers and brands looking to optimize their influencer marketing strategies. Here are some ways ChatGPT can be used for influencer marketing:

1. Finding the Right Influencers: ChatGPT can be used to identify the best influencers for a brand by analyzing factors like engagement rates, audience demographics, and previous brand collaborations.
2. Generating Content Ideas: ChatGPT can be used to generate content ideas for influencer campaigns, such as caption ideas, post themes, and hashtags.
3. Optimizing Hashtags: ChatGPT can be used to identify the best hashtags to use in influencer campaigns by analyzing the most popular hashtags in a given niche or industry.
4. Analyzing Campaign Performance: ChatGPT can be used to analyze the performance of influencer campaigns by tracking engagement rates, click-through rates, and conversion rates.
5. Managing Influencer Relationships: ChatGPT can be used to manage influencer relationships by tracking communication, scheduling posts, and monitoring campaign performance.

Identifying Potential Collaborations with AI

ChatGPT can be used to identify potential collaborations with influencers. By analyzing the content, engagement, and followers of different influencers, ChatGPT can suggest potential partnerships that align with a brand's goals and target audience. Additionally, ChatGPT can help create a list of influencers and provide insights into their follower demographics and interests to determine if they would be a good fit for a particular campaign. This can save time and resources in the influencer outreach process and help ensure a successful partnership.

Managing Influencer Relationships with ChatGPT

ChatGPT can help social media managers to manage influencer relationships in various ways.

Firstly, ChatGPT can help to identify the right influencers to collaborate with. Social media managers can use ChatGPT to analyze data on influencers, such as their followers, engagement rates, and demographics, and identify those who are likely to be a good fit for their brand.

Secondly, ChatGPT can help social media managers to communicate with influencers. ChatGPT can be used to automate and personalize messages, such as outreach emails, and to manage ongoing communication with influencers.

Thirdly, ChatGPT can be used to measure the impact of influencer collaborations. Social media managers can use ChatGPT to analyze data on engagement rates, conversions, and other metrics, and to determine the ROI of influencer collaborations.

Finally, ChatGPT can be used to develop and implement influencer marketing strategies. Social media managers can use ChatGPT to analyze data on trends and consumer behavior, and to develop strategies that align with these trends. ChatGPT can also be used to track the success of these strategies and to make adjustments as needed.

Chapter 9:
Ethical Considerations in AI-Assisted Social Media Management

Social media managers need to be aware of ethical considerations when using AI for social media management. Here are some important ethical considerations:

1. Privacy: When using AI tools to analyze social media data, it is important to ensure that the privacy of individuals is respected. Personal data should be kept confidential, and only used for the purpose for which it was collected.
2. Bias: AI tools can be biased, based on the data they are trained on. Social media managers need to be aware of this, and work to ensure that AI tools are not discriminating against certain groups of people.
3. Transparency: Social media managers need to be transparent about their use of AI tools. They should inform their audience when AI tools are being used, and how they are being used.
4. Responsibility: Social media managers need to take responsibility for the actions of their AI tools. They should ensure that their tools are not causing harm, and that they are being used in an ethical manner.
5. Fairness: AI tools should be used fairly, and not to gain an unfair advantage over others. Social media managers need to ensure that their use of AI tools does not violate ethical standards or harm others.

By being aware of these ethical considerations, social media managers can use AI tools in a responsible and ethical manner.

Responsible Use of AI on Social Media

As with any technology, it is important to use AI in social media management responsibly and ethically. Here are some considerations:

1. Transparency: It is important to be transparent about the use of AI in social media management. Users should be aware that they are interacting with AI-powered chatbots, for example.
2. Bias: AI algorithms are only as good as the data they are trained on, so it is important to ensure that the data is representative and unbiased.
3. Privacy: Social media platforms contain sensitive personal information, and it is important to protect users' privacy when using AI.

4. Human oversight: While AI can be a powerful tool, it should not replace human oversight entirely. Human judgement is still needed to ensure that AI is being used in a responsible and ethical way.
5. Accessibility: AI should be accessible to all users, regardless of their abilities or background. Designing AI-powered tools with accessibility in mind can ensure that everyone can benefit from them.
6. Accountability: Social media managers should be accountable for the decisions made by AI-powered tools. Regular audits and reviews can help ensure that AI is being used responsibly and effectively.

Privacy, Transparency, and Trust in AI Tools

As the use of AI becomes more prevalent in social media management and other fields, it is important to consider the ethical implications of these tools. One key issue is privacy, as AI algorithms may be used to analyze user data without their knowledge or consent. It is important for social media managers to be transparent about their use of AI and to ensure that user data is protected and used responsibly.

Another issue is transparency, as some AI algorithms may be opaque and difficult to understand. Social media managers should seek to use tools that are explainable and can be audited, so that they can understand how the algorithm is making decisions and ensure that it is not biased.

Finally, trust is a crucial factor in the use of AI tools. Social media managers should be aware of the potential for AI to perpetuate biases or discrimination, and should take steps to ensure that their use of AI is fair and unbiased. This may include using diverse data sets and monitoring the performance of the algorithm over time.

Overall, social media managers should be aware of the ethical considerations surrounding the use of AI in their work, and should take steps to ensure that their use of AI is responsible, transparent, and ethical.

Chapter 10:
Looking Ahead: ChatGPT and the Future of Social Media Management

As social media continues to play a significant role in business and personal communication, the use of AI is likely to become even more prevalent in social media management. Some current trends and future predictions include:

1. Personalized content: With the help of AI, social media managers will be able to generate personalized content that resonates with individual users based on their interests, behavior, and engagement history.
2. Chatbots and automation: The use of chatbots and automation will become increasingly widespread in social media management. AI-powered chatbots can quickly and accurately respond to customer queries, complaints, and feedback, helping to improve customer satisfaction and retention.
3. Predictive analytics: Predictive analytics will play a critical role in social media management, allowing businesses to anticipate trends, identify opportunities, and make informed decisions.
4. Social listening: Social listening tools that use AI will become more sophisticated, allowing businesses to monitor and analyze conversations on social media in real-time, and take action accordingly.
5. Augmented reality: Augmented reality (AR) is likely to become more common in social media, allowing businesses to create interactive and immersive experiences for users.
6. Ethical considerations: As AI becomes more prevalent in social media management, there will be a greater need for ethical considerations and responsible use of these tools. Privacy, transparency, and trust will be critical factors in the adoption and use of AI-powered social media tools.

Overall, AI is set to revolutionize social media management, helping businesses to create more engaging content, improve customer satisfaction, and make more informed decisions. As the technology continues to evolve, it is likely that we will see even more innovative and exciting uses for AI in social media management.

Current Trends and Future Predictions

Social media continues to evolve at a rapid pace, and AI is expected to play an increasingly significant role in social media management in the coming years. Some current trends and future predictions include:

1. Greater automation: AI tools will continue to automate routine tasks such as content scheduling, monitoring and responding to messages, and analyzing engagement metrics.
2. Personalization: AI will enable social media managers to deliver more personalized content to specific audiences, improving engagement and increasing the likelihood of conversion.
3. Improved analytics: AI-powered analytics tools will become more sophisticated, enabling managers to gain deeper insights into audience behavior and sentiment.
4. Chatbots and voice assistants: As voice search and chatbots become more prevalent, social media managers will need to adapt their content to be discoverable via these channels.
5. Augmented reality: AR is expected to play a more significant role in social media marketing, enabling users to try out products virtually before making a purchase.
6. Enhanced security: As the risk of data breaches and cyberattacks continues to rise, AI will play a greater role in ensuring the security and privacy of social media data.

Overall, AI is expected to revolutionize social media management, enabling managers to be more efficient, effective, and responsive to their audiences. However, as with any emerging technology, it is important to consider the ethical implications of AI-powered social media tools and ensure that they are used in a responsible and transparent manner.

Embracing AI in the Social Media Landscape

As the volume of data generated by social media continues to increase, the role of AI in social media management will become even more important. AI tools can help social media managers make sense of this data, identify trends, and gain insights that would be difficult to discern manually. AI can also help automate routine tasks, such as post scheduling and engagement monitoring, allowing social media managers to focus on creating more meaningful content and building relationships with their audience.

In the future, it's likely that AI will become even more integral to social media management. As natural language processing and machine learning algorithms improve, AI tools will be better equipped to analyze and understand the nuances of human communication on social media platforms. This could lead to more sophisticated sentiment analysis, improved chatbot interactions, and even the development of AI-generated content that is virtually indistinguishable from content created by humans.

However, with the growing use of AI in social media management, it's important to be aware of potential ethical considerations, such as data privacy and algorithmic bias. It will be crucial for social media managers to ensure that they are using AI tools in a responsible and transparent manner, and that they are taking steps to mitigate any potential negative impacts on their audience or the wider community.

The best PROMPTs for Social Media Manager:

Write in GPT Chat this prompt exactly as it is written below. Then try changing the terms you find in the " " to get the work that works best for you. Remember that in case you are a social media manager, this prompt is PERFECT for you, it is a very powerful tool make good use of it! Prompt:

You are a social media manager and will act like one. Keeping in mind that I have a business based on "women's clothing", the brand name is "Fashion Dress" create a daily publication plan lasting one month, then organize 30 posts on social networks "Facebook and Instagram". This publication plan should be as long and detailed as possible and will serve to retain customers and convert them. One post per week will have to aim to go viral. You will format the article in markdown. You will use an "analytical" writing style, a "formal" tone, and a "professional" communicative register.

BOOK 9: "ChatGPT for Journalists: Research, Writing and Fact-checking": Examines how ChatGPT can be used in the field of journalism.

Chapter 1:
Introduction to ChatGPT for Journalists

As a language model trained by OpenAI, ChatGPT can assist journalists in several ways, including generating ideas for articles, providing research assistance, fact-checking, and more. In this section, we will discuss how ChatGPT can be used by journalists to improve their workflow and enhance their reporting.

Journalism is a field that requires creativity, critical thinking, and excellent writing skills. However, the advent of digital technology has changed the landscape of journalism, leading to an increase in the amount of content being produced, shorter deadlines, and more emphasis on clicks and shares. As a result, journalists are constantly looking for ways to improve their productivity and efficiency without sacrificing the quality of their work. This is where AI tools like ChatGPT come in handy, helping journalists to streamline their work, improve their research and writing, and ultimately produce better stories.

In the following sections, we will explore the various ways in which journalists can use ChatGPT to enhance their work.

Understanding ChatGPT

ChatGPT is a large language model trained by OpenAI that is capable of generating human-like responses to text prompts. It is based on the GPT (Generative Pretrained Transformer) architecture, which allows it to understand and generate natural language text in a wide range of contexts. ChatGPT is trained on a massive amount of data from the internet, which allows it to learn and replicate patterns of language and writing. As a result, it has a range of potential applications in journalism, including assisting with research, generating story ideas, and automating certain aspects of the writing process.

The Role of AI in Journalism

The role of AI in journalism is rapidly growing and evolving. AI can be used to automate tasks that were previously done manually, such as fact-checking, transcribing interviews, and data analysis. It can also be used to generate news stories, summarize articles, and identify patterns and trends in large datasets. AI can help journalists save time, increase accuracy, and enhance their ability to identify and cover important stories. However, there are also concerns about the impact of AI on the future of journalism, such as the potential for bias and the potential loss of jobs as more tasks become automated.

Chapter 2:
Research and Investigation with ChatGPT

ChatGPT can be an invaluable tool for journalists during the research and investigation phase of their work. AI models such as ChatGPT can process and analyze large amounts of data, helping journalists to quickly identify key information and trends.

ChatGPT can also help journalists to identify potential sources and experts to contact for interviews or quotes. By analyzing online content, ChatGPT can identify individuals who have expertise in a particular area, and can provide journalists with their contact information.

Additionally, ChatGPT can help journalists to fact-check their work. By analyzing sources and comparing information, ChatGPT can help journalists identify inconsistencies or inaccuracies in their reporting. This can help to ensure that journalists are producing accurate and reliable content.

However, it's important to note that AI tools like ChatGPT should be used as a complement to, rather than a replacement for, traditional journalistic skills and practices. Journalists must still conduct their own research, verify sources, and fact-check their work to ensure that it meets the highest standards of accuracy and integrity.

Streamlining the Research Process with AI

ChatGPT can be a valuable tool for journalists in streamlining the research process. With its ability to sift through large amounts of data quickly and efficiently, it can help journalists to identify patterns and connections that might otherwise go unnoticed. For example, journalists can use ChatGPT to quickly scan through large volumes of government documents or public records to find relevant information for their stories.

Additionally, ChatGPT can help journalists to identify and verify sources. With its ability to analyze text and language, it can assist in identifying patterns of language use that may indicate a source's authenticity or reliability. It can also help to identify potential conflicts of interest or other factors that may impact the credibility of a source.

Overall, ChatGPT can help journalists to conduct more thorough and effective research, ultimately leading to more accurate and informative reporting.

Digging Deeper with Data Analysis

With the amount of data available today, journalists have access to a wealth of information that can help them to uncover stories and trends that might have otherwise gone unnoticed. However, this data is often complex and difficult to analyze without the help of specialized tools. This is where AI and ChatGPT can be particularly useful.

With AI-powered data analysis, journalists can quickly sift through vast amounts of data to identify patterns, correlations, and anomalies. This can help them to uncover new angles on stories, identify potential sources, and make connections between different pieces of information. For example, AI-powered analysis could be used to comb through financial records to identify suspicious transactions, or to analyze social media activity to identify emerging trends or issues.

ChatGPT can also be used to help journalists with more basic research tasks. For example, it can be used to quickly generate background information on a particular topic, to find relevant sources, or to identify key themes or topics that might be relevant to a particular story. This can help to save journalists time and allow them to focus on the more important aspects of their work.

Overall, AI and ChatGPT have the potential to greatly enhance the research and investigation capabilities of journalists, allowing them to uncover new stories and insights more quickly and efficiently.

Chapter 3:
Writing and Editing News Stories with ChatGPT

ChatGPT can be a useful tool for journalists in writing and editing news stories. Here are some ways ChatGPT can assist in the process:

1. Generating article summaries: ChatGPT can analyze a news story and provide a concise summary of its main points. This can be useful for journalists who need to quickly review articles for research or fact-checking.
2. Writing headlines: ChatGPT can suggest headlines based on the content of a news story. This can save journalists time in coming up with compelling headlines that accurately reflect the story.
3. Fact-checking: ChatGPT can be used to fact-check news stories and ensure accuracy. By inputting a claim or statement, ChatGPT can quickly analyze data and provide information to support or refute the claim.
4. Writing assistance: ChatGPT can provide suggestions for sentence structure, grammar, and vocabulary in news stories. This can be helpful for journalists who are working under tight deadlines or who need assistance with their writing.
5. Plagiarism checking: ChatGPT can be used to check for plagiarism in news stories. By comparing a journalist's writing to a vast database of articles, ChatGPT can flag any potential instances of plagiarism.
6. Editing and proofreading: ChatGPT can assist in editing and proofreading news stories. By analyzing the text, ChatGPT can suggest changes to improve clarity, conciseness, and overall quality.

Drafting Articles with AI Assistance

ChatGPT can be a useful tool for journalists in drafting news articles with AI assistance. Journalists can use the AI model to generate initial ideas, refine story angles, and suggest relevant sources for interviews. They can also use it to help develop an outline and structure for their articles. ChatGPT can also assist in identifying potential bias and improving the accuracy of news stories.

It is important to note that AI should not be a substitute for the journalistic skills of researching, verifying sources, and fact-checking. Journalists should always verify any information generated by AI before publishing it as news. The AI model should be used as a tool to assist journalists in their work, not as a replacement for their expertise and critical thinking.

Editing and Proofreading with ChatGPT

Editing and proofreading are two crucial steps in the journalism process, and ChatGPT can be a valuable tool in both areas. Here are some ways that ChatGPT can assist in editing and proofreading:

1. Grammar and spelling: ChatGPT can be used to check the grammar and spelling of articles. It can identify common errors such as misspellings, incorrect punctuation, and subject-verb agreement issues.
2. Style and consistency: ChatGPT can help to ensure that articles are consistent in terms of style and tone. It can flag any inconsistencies and suggest changes to ensure that the article reads smoothly.
3. Fact-checking: ChatGPT can help to fact-check articles by identifying any factual errors or inconsistencies. It can also suggest sources for further research to ensure that articles are accurate.
4. Clarity and readability: ChatGPT can help to improve the clarity and readability of articles. It can identify areas where the language is overly complex and suggest simpler alternatives. It can also flag any areas where the article may be difficult to understand for the target audience.
5. Plagiarism detection: ChatGPT can be used to check articles for plagiarism. It can compare the article to other sources and flag any areas where the language is similar.

It is important to note that while ChatGPT can be a helpful tool in editing and proofreading, it should not be relied upon as the sole means of ensuring the quality and accuracy of articles. Human editors and proofreaders are still essential in the journalism process to ensure that articles are thoroughly reviewed and edited.

Chapter 4:
Fact-checking with ChatGPT

Fact-checking is a critical component of journalism, and AI can play a role in helping journalists to verify the accuracy of the information they report. ChatGPT can be used to fact-check information by cross-referencing it with credible sources and identifying any inconsistencies or errors. For example, a journalist could input a claim made by a source, and ChatGPT could analyze the statement, searching for corroborating evidence or discrepancies in the information presented. AI-powered fact-checking can help journalists to work more efficiently and accurately, but it's important to keep in mind that AI is not infallible and should always be used in conjunction with human expertise and critical thinking skills.

Using AI to Validate Information

ChatGPT can be used to validate information and fact-check articles. With its language processing capabilities, it can analyze text and cross-reference it with trusted sources to identify any inaccuracies or errors.

For example, a journalist could input a statement or claim into ChatGPT and ask it to verify its accuracy by cross-referencing it with reputable sources. ChatGPT could then provide a list of sources that either support or contradict the statement, allowing the journalist to assess the accuracy of the information and make any necessary corrections.

In addition, ChatGPT could be used to identify and flag potential instances of bias or propaganda in news articles, helping journalists to maintain objectivity and integrity in their reporting.

Spotting Fake News and Misinformation

Identifying Misinformation and Fake News with ChatGPT

As a language model, ChatGPT has the ability to analyze text and identify patterns that may indicate the presence of misinformation or fake news. Here are some ways ChatGPT can help in identifying misinformation and fake news:

1. Check sources: ChatGPT can analyze the source of a news story or article and compare it to known reliable sources to determine its credibility.
2. Analyze content: ChatGPT can analyze the content of a news story or article to identify any

claims that may be false or misleading.

3. Fact-checking: ChatGPT can compare statements made in a news story or article to known facts and statistics to determine their accuracy.
4. Cross-checking: ChatGPT can cross-check information from multiple sources to determine its accuracy and identify inconsistencies.
5. Identify bias: ChatGPT can analyze the language and tone of a news story or article to determine any underlying bias or agenda.

It is important to note that ChatGPT is only a tool and should be used in conjunction with human judgement and critical thinking when assessing the credibility of news stories and articles.

Chapter 5:
Interviewing and Reporting with ChatGPT

ChatGPT can be a valuable tool for journalists in conducting interviews and gathering information for their stories. Here are some ways that ChatGPT can assist with interviewing and reporting:

1. Researching interviewees: ChatGPT can help journalists learn more about the people they are interviewing by providing background information on their education, work experience, and other relevant details. This can help journalists come up with better questions and conduct more informed interviews.
2. Generating interview questions: ChatGPT can assist journalists in generating interview questions by suggesting relevant topics and follow-up questions. This can help journalists cover all the necessary ground and ensure they don't miss any important details.
3. Conducting virtual interviews: With the rise of remote work, many interviews are now conducted online or over the phone. ChatGPT can assist with this by providing suggestions for how to phrase questions or respond to answers in a virtual context.
4. Transcription: ChatGPT can assist with transcription of interviews, saving journalists time and allowing them to focus on analyzing the content of the interviews.
5. Translation: ChatGPT can also be used to translate interviews conducted in other languages into English or another language the journalist is more comfortable with.
6. Reporting assistance: ChatGPT can assist journalists in generating summaries of information and even write news reports using the information gathered from interviews.

Overall, ChatGPT can be a valuable tool for journalists in conducting interviews and reporting the news. However, it is important for journalists to use AI tools responsibly and critically evaluate the information provided by AI to ensure accuracy and fairness in their reporting.

Preparing for Interviews with AI

Preparing for interviews with AI can be an effective way for journalists to save time and increase their chances of conducting a successful interview. ChatGPT can assist journalists in various ways, such as generating interview questions, researching the interviewee's background, and identifying potential areas of interest.

To prepare for an interview with ChatGPT, journalists can start by researching the interviewee's background, including their professional history and any relevant publications or projects. They can then generate a list of potential questions based on this information, using ChatGPT to help refine and expand the list as needed.

During the interview, journalists can use ChatGPT to transcribe the conversation and highlight key points or topics of interest. They can also use AI to analyze the interviewee's tone and body language to help identify potential areas of follow-up or additional inquiry.

After the interview, ChatGPT can assist in organizing and summarizing the information gathered, as well as identifying potential angles for a news story or feature. Overall, AI can help streamline the interview process and provide valuable insights for journalists.

Live Reporting and Transcribing with ChatGPT

ChatGPT can assist in live reporting and transcribing in a variety of ways. It can be used to transcribe live events such as press conferences, speeches, and interviews, providing real-time transcripts that can be used for reporting. ChatGPT can also be used to help journalists keep track of multiple sources of information simultaneously, by monitoring social media feeds, live streams, and other sources of data and providing updates and alerts in real-time.

In addition to live reporting, ChatGPT can also assist with transcribing pre-recorded interviews and other audio or video content. This can save journalists a significant amount of time and effort, allowing them to focus on the actual reporting and analysis of the content, rather than the transcription process.

ChatGPT can also be used to generate summaries of longer content, such as reports or research papers, allowing journalists to quickly digest and understand complex information. This can be especially useful when working on tight deadlines or when covering complex subjects that require a significant amount of research and analysis.

Chapter 6:
Data Journalism and Visualization with ChatGPT

Data journalism involves collecting, analyzing, and presenting data in a way that tells a story or uncovers trends or patterns. ChatGPT can be used to assist journalists in these tasks, making the process faster and more efficient.

Data Analysis with ChatGPT: ChatGPT can be used to analyze large datasets, such as government data, social media data, and other publicly available data. Journalists can use ChatGPT to quickly identify patterns and trends in the data, allowing them to tell more compelling stories.

Data Visualization with ChatGPT: ChatGPT can also be used to create visualizations of data, such as charts, graphs, and other interactive graphics. These visualizations can help readers understand complex data sets more easily and can add depth and context to a story.

Fact-checking with ChatGPT: In addition to analyzing and visualizing data, ChatGPT can be used to fact-check information, including statistics and claims made in speeches and other public statements. Journalists can use ChatGPT to quickly verify information and ensure that their reporting is accurate.

Story Generation with ChatGPT: Finally, ChatGPT can also be used to generate new story ideas based on data trends and patterns. Journalists can use ChatGPT to identify interesting topics and angles that they may have missed otherwise.

Overall, ChatGPT can be a powerful tool for data journalists, allowing them to uncover new information, verify facts, and tell more compelling stories. However, it is important to use AI tools responsibly and to verify the accuracy of their outputs.

Analyzing Large Data Sets with AI

ChatGPT can be used for analyzing large data sets in data journalism. With its ability to understand and process natural language, it can help journalists extract relevant information and insights from vast amounts of data. For example, ChatGPT can be used to analyze political speeches, social media trends, and public records to identify patterns and trends that are relevant to a particular story.

ChatGPT can also be used to create data visualizations, such as charts and graphs, which can help readers understand complex data sets more easily. By using natural language queries to identify the most important data points and trends, ChatGPT can make it easier for journalists to create compelling visualizations that effectively communicate their findings to their audience.

In addition, ChatGPT can be used for data mining, which involves searching through large data sets to identify patterns and insights that might not be immediately apparent. This can be particularly useful in investigative journalism, where journalists may need to uncover hidden connections or relationships between people, organizations, or events. With its ability to analyze natural language text and understand the context of words and phrases, ChatGPT can help journalists identify important leads and piece together complex stories.

Creating Data Visualizations with ChatGPT

ChatGPT can assist in creating data visualizations by providing recommendations for the type of chart, graph, or visualization that is most appropriate for the data being presented. It can also help with labeling and formatting of the visualizations. Additionally, ChatGPT can provide suggestions for presenting data in a way that is both informative and engaging for the audience. However, it is important to note that ChatGPT is not a data analysis tool and should not be relied on solely for making data-driven decisions. It is important to verify the accuracy of data and ensure that the visualizations are appropriately labeled and interpreted.

Chapter 7:
Social Media and Citizen Journalism

Social media has revolutionized the way people consume news and information, and citizen journalism has emerged as a powerful force in the field of journalism. Citizen journalism refers to the practice of non-professional individuals engaging in the gathering, reporting, and dissemination of news and information. Social media platforms such as Twitter, Facebook, and YouTube have made it easier than ever for citizens to share news and information with the world.

ChatGPT can play a role in monitoring social media trends and engaging with the public and citizen journalists. With its ability to analyze large data sets and identify patterns and trends, ChatGPT can help journalists stay on top of breaking news and emerging trends. It can also assist with fact-checking and identifying sources.

In addition, ChatGPT can help journalists engage with the public and citizen journalists by providing a platform for discussion and feedback. Journalists can use ChatGPT to ask questions, solicit feedback, and interact with their audience in real-time. This can help build trust and foster a sense of community around the news and information being shared.

Finally, ChatGPT can be used to identify and track social media influencers and opinion leaders. By analyzing social media data and identifying patterns of influence, journalists can better understand the social dynamics at play in the online conversation and craft stories that resonate with their audience.

Monitoring Social Media Trends with ChatGPT

ChatGPT can be a powerful tool for monitoring social media trends and conversations. With its natural language processing capabilities, it can quickly sift through vast amounts of social media data and identify relevant trends and topics.

One way to use ChatGPT for social media monitoring is by setting up a custom search query that targets specific keywords or hashtags related to your industry or brand. ChatGPT can then analyze the resulting social media posts and provide insights into how people are talking about those topics.

Another approach is to use ChatGPT to track sentiment analysis, which involves analyzing the language and tone used in social media posts to determine whether they are positive, negative, or neutral. This can help you gauge how your brand is being perceived online and identify potential issues that need to be addressed.

Overall, ChatGPT can be a valuable tool for journalists looking to stay on top of the latest social media trends and monitor conversations relevant to their reporting.

Engaging with the Public and Citizen Journalists

ChatGPT can assist journalists in engaging with the public and citizen journalists in a number of ways. For example:

1. Generating article ideas: ChatGPT can help journalists generate new and unique story ideas by analyzing trending topics, public sentiment, and other factors.
2. Fact-checking: ChatGPT can be used to quickly fact-check information from social media or other sources, allowing journalists to verify the accuracy of a story before publishing.
3. Identifying sources: ChatGPT can help journalists identify potential sources for a story, based on their expertise, location, or other factors.
4. Social media monitoring: ChatGPT can monitor social media platforms for mentions of a particular topic, brand, or event, allowing journalists to stay on top of breaking news and public sentiment.
5. Translation: ChatGPT can be used to translate social media posts or other content from different languages, making it easier for journalists to follow international events and trends.

Overall, ChatGPT can help journalists stay on top of emerging trends and breaking news, connect with new sources, and ensure that their reporting is accurate and well-informed.

Chapter 8:
Crisis Reporting and Risk Management

Crisis reporting and risk management are critical components of journalism that require careful planning, quick thinking, and a willingness to adapt to new situations. With the help of AI, journalists can streamline their reporting process and access vital information more efficiently during times of crisis.

One way that AI can assist in crisis reporting is by analyzing social media and other online sources for breaking news and real-time updates. ChatGPT can be programmed to monitor specific keywords and phrases on social media platforms, news sites, and other sources of information, alerting journalists to relevant developments as they happen. This can be particularly useful during natural disasters, political upheavals, and other crises where up-to-date information is essential.

AI can also assist in risk management by identifying potential threats and vulnerabilities that journalists may face while reporting. For example, ChatGPT can analyze news reports, social media activity, and other sources of information to identify dangerous areas, volatile situations, and other risks that journalists may face. By providing this information in real-time, AI can help journalists make informed decisions about where and how to report on a story, reducing the likelihood of harm or injury.

Finally, AI can assist in crisis reporting by helping journalists manage large volumes of data and information. During a crisis, journalists may need to sort through vast amounts of data, reports, and other materials to piece together a cohesive narrative. AI can help automate this process, analyzing and organizing data in a way that allows journalists to quickly identify trends, patterns, and other key information.

While AI can be a valuable tool in crisis reporting and risk management, it's important for journalists to use these tools responsibly and ethically. Journalists must always verify their sources, fact-check information, and ensure that their reporting is accurate, fair, and unbiased. Additionally, journalists should be transparent about their use of AI and the role it plays in their reporting, helping to build trust with their audiences and maintain the integrity of their work.

Using ChatGPT for Rapid Response Reporting

ChatGPT can be useful for rapid response reporting during a crisis. Journalists can use AI to monitor social media and other sources to quickly identify potential breaking news stories. ChatGPT can be used to analyze data, identify patterns, and help journalists spot trends and potential leads. AI-powered chatbots can also help journalists answer common questions and provide information to the public during a crisis. Additionally, AI can help with risk management by providing journalists with information about potential safety risks and providing recommendations for staying safe while reporting. Overall, ChatGPT can be a powerful tool for journalists who need to respond quickly and accurately to a breaking news event or crisis.

Assessing Risks and Safety in Conflict Zones

As an AI language model, I cannot assess risks or provide advice on safety measures in conflict zones as this requires specific expertise and knowledge. It is important to seek guidance from professional journalists and experts in conflict reporting who have experience working in these environments. They can provide information on best practices, risk assessments, and safety measures to ensure the well-being of reporters and the accuracy and credibility of the information being reported. Additionally, there are organizations that specialize in training journalists for working in high-risk areas, and they can offer guidance and resources on this topic.

Chapter 9:
Ethical Considerations in AI-Assisted Journalism

As with any use of artificial intelligence, there are ethical considerations to keep in mind when using ChatGPT for journalism. Here are some key areas to consider:

1. Bias: AI models are only as good as the data they are trained on, so if the training data is biased, the AI may also be biased. It's important to ensure that the data used to train the model is diverse and representative of different perspectives and experiences.
2. Privacy: When using AI to analyze data or gather information, it's important to consider privacy concerns. Make sure you are following appropriate data protection regulations and are transparent with your audience about how their data is being used.
3. Accuracy: While AI can be a powerful tool for analyzing and summarizing information, it is not infallible. It's important to fact-check and verify any information before publishing it, especially when using AI to gather or analyze data.
4. Transparency: When using AI to generate content or assist in reporting, it's important to disclose that fact to your audience. This can help build trust and ensure that your reporting is seen as authentic and trustworthy.
5. Responsibility: Ultimately, journalists are responsible for the content they publish, regardless of whether AI was used in the process. It's important to consider the impact of your reporting and ensure that it is fair, accurate, and respectful of all parties involved.

By keeping these considerations in mind and using AI tools responsibly, journalists can harness the power of technology to enhance their reporting and reach new audiences.

Responsible Use of AI in Journalism

As with any technology, the use of AI in journalism requires ethical considerations to ensure that its use does not compromise the integrity of journalism. Here are some key ethical considerations for using AI in journalism:

1. Accuracy: AI-generated content should be accurate, unbiased, and factually correct. Journalists should not rely solely on AI-generated content without verifying its accuracy and authenticity.
2. Transparency: Journalists should disclose when AI is used in generating content, such as by including an AI-generated disclaimer or tag. Transparency builds trust between journalists and

their audiences.
3. Diversity: AI-generated content should be inclusive and represent diverse perspectives. Journalists should avoid using AI in a way that perpetuates stereotypes or biases.
4. Accountability: Journalists should take responsibility for the AI-generated content they produce and ensure that it is in line with journalistic standards and ethics.
5. Human judgment: While AI can assist journalists in the gathering and processing of information, human judgment should still be prioritized in determining what stories to cover and how they should be presented.
6. Privacy: Journalists should be mindful of the privacy concerns that arise when using AI for data collection and analysis. They should ensure that any personal information is collected and used in compliance with data protection regulations.

Overall, the responsible use of AI in journalism requires a balance between the efficiency and accuracy that AI can provide and the ethical considerations that must be taken into account.

Addressing Bias, Privacy, and Ethics in AI Tools

Addressing bias, privacy, and ethics in AI tools is essential to ensure responsible use and protect the public's rights and interests. AI tools like ChatGPT are designed to learn from large amounts of data and make decisions based on that data. However, these tools may reflect the biases and prejudices present in the data, which can lead to discrimination and injustice. Therefore, it is crucial to take steps to identify and correct any biases in AI tools, particularly in applications that impact society's most vulnerable groups, such as in journalism.

Privacy is another critical issue in AI-assisted journalism. AI tools like ChatGPT rely on large datasets to learn and make decisions, often containing sensitive personal information. Journalists must ensure that any data used in AI-assisted journalism is collected and stored securely, and any personal information is anonymized to protect individuals' privacy.

Ethical considerations must also be taken into account when using AI tools in journalism. For example, journalists must ensure that they do not use AI to generate fake news or manipulate public opinion. They must also be transparent about their use of AI tools in their reporting, particularly when generating content. Additionally, journalists should ensure that any AI-assisted content is reviewed and edited by human editors to maintain the highest standards of accuracy and quality.

Overall, AI tools like ChatGPT have the potential to transform the field of journalism, making it faster, more efficient, and more accessible. However, it is essential to address any potential biases, privacy concerns, and ethical considerations to ensure responsible and ethical use of these tools.

Chapter 10:
Looking Ahead: ChatGPT and the Future of Journalism

As an AI language model, ChatGPT cannot predict the future with certainty. However, it is clear that AI technology will continue to have a significant impact on the field of journalism. Some potential future developments include:

1. More advanced AI tools for research and investigation, allowing journalists to uncover new stories and find deeper insights from data.
2. Increased use of AI for fact-checking and verification, helping to combat the spread of fake news and misinformation.
3. Greater use of AI for automated content creation, such as news articles and summaries, freeing up journalists to focus on more in-depth reporting.
4. The continued rise of citizen journalism and social media, creating new challenges and opportunities for journalists to engage with their audience and report on breaking news in real-time.
5. Further discussions and debates about the ethical implications of using AI in journalism, including issues such as bias, privacy, and accountability.

Overall, AI technology is likely to play an increasingly important role in journalism, enabling journalists to work more efficiently, access new sources of information, and provide their audiences with more accurate and engaging content. However, as with any new technology, it will be important for journalists to approach AI with a critical eye, and to consider the potential risks and challenges associated with its use.

Current Trends and Future Predictions

There are several current trends and future predictions for AI in journalism:

1. Data-driven journalism: With the increasing availability of data and the development of AI tools to analyze it, data-driven journalism is becoming more popular. AI can help journalists identify patterns and trends that may not be immediately apparent in large data sets, leading to more accurate reporting.
2. Automated news writing: AI is already being used to generate news stories, and this trend is likely to continue. While AI-generated content may lack the nuance and creativity of human-

written articles, it can be useful for producing news stories quickly and efficiently.

3. Fact-checking: AI-powered fact-checking tools are becoming more sophisticated and are being used by news organizations to verify information quickly and accurately.
4. Personalization: News organizations are using AI to personalize content for individual users. This can help to improve engagement and loyalty, but it also raises concerns about the echo chamber effect.
5. Chatbots: AI-powered chatbots are being used by news organizations to interact with readers and provide them with personalized news and information.
6. Virtual and augmented reality: VR and AR technologies are being used by news organizations to provide immersive experiences for readers. This trend is likely to continue as the technology becomes more advanced and accessible.
7. Ethics and transparency: As AI becomes more prevalent in journalism, there is a growing need for transparency and ethical guidelines around its use. News organizations will need to ensure that AI is being used in a responsible and unbiased way.

Overall, AI is likely to play an increasingly important role in journalism in the coming years. While there are concerns about the potential for bias and inaccuracies, the benefits of AI in terms of speed, accuracy, and efficiency are hard to ignore.

Preparing for an AI-Driven Journalism Landscape

As artificial intelligence (AI) continues to evolve and permeate different aspects of our lives, it is increasingly becoming an important tool in the field of journalism. With its ability to analyze and process large amounts of data, automate routine tasks, and provide insights and recommendations, AI is transforming the way journalists work and report on news.

One of the major trends in AI-driven journalism is the use of chatbots and natural language processing (NLP) to create more engaging and personalized experiences for readers. Chatbots can be used to deliver news stories, provide recommendations based on users' interests, and even engage in conversations with readers. NLP, on the other hand, can be used to extract insights from large volumes of text data, summarize news articles, and even generate new content.

Another trend is the use of data-driven journalism, which involves using AI to analyze and visualize data to identify trends, patterns, and insights. This can help journalists uncover hidden stories, find new angles, and present data in a more understandable and accessible way.

AI is also being used to automate certain tasks, such as transcription, translation, and fact-checking, which can help journalists save time and focus on more important aspects of their work. Additionally, AI can be used to monitor social media and other online platforms for breaking news and user-generated content, allowing journalists to stay up-to-date and get a more complete picture of events as they unfold.

As AI continues to advance, it is likely that it will play an even greater role in journalism in the future. However, it is important for journalists to approach this technology with caution and ensure that they use it responsibly, taking into account issues such as bias, privacy, and ethics. Ultimately, AI should be seen as a tool to augment and enhance journalistic work, rather than replace it.

The best PROMPTs for Journalists:

Write in GPT Chat this prompt exactly as it is written below. Then try changing the terms you find in the " " to get the work that works best for you. Remember that in case you are a journalist, this prompt is PERFECT for you, it is a very powerful tool make good use of it! Prompt:

You are a journalist and will act like one. You will write an "informative" themed article entitled "America's Top Cities," intended for an audience of "men, over 50," with the goal of "increasing the brand reputation of these cities." The article will have the following structure: Title, introduction, 1 H2 section, 1 H3 section. The H3 sections will be subsections of the H2 sections. You will format the article in markdown. You will use a "professional" writing style, an "engaging" sentiment, and an "informal" communicative register.

BOOK 10: "ChatGPT for Linguists: Translation, Interpretation and Language Learning": Discusses the use of ChatGPT for language learning and translation.

Chapter 1:
Introduction to ChatGPT for Linguists

As an AI language model, I can provide an overview of ChatGPT for linguists. ChatGPT is a large-scale deep learning language model developed by OpenAI that can generate text in a variety of styles and formats, such as news articles, essays, and social media posts. Linguists can use ChatGPT to explore language patterns and generate natural language text for various research and practical applications.

ChatGPT is trained on a massive dataset of text from the internet, including books, articles, and other sources. It uses deep learning techniques to learn the statistical patterns and relationships between words and sentences, and then generates text that is similar to the input data.

Linguists can use ChatGPT for a wide range of applications, such as analyzing language patterns, generating new language data for research, and creating language models for various language processing tasks. With its ability to generate text in a natural and coherent manner, ChatGPT has the potential to transform the field of linguistics and language processing.

Some potential use cases for ChatGPT in linguistics include:

- Language analysis: Linguists can use ChatGPT to analyze language patterns and explore the relationship between words and syntax.
- Language modeling: ChatGPT can be used to generate language models that can be used in natural language processing tasks such as language translation, sentiment analysis, and speech recognition.
- Language generation: Linguists can use ChatGPT to generate new language data for research and analysis. This can be particularly useful for low-resource languages or languages that lack adequate language resources.
- Language teaching: ChatGPT can be used to generate language exercises and quizzes for language learners, and can also be used to develop interactive language learning tools and applications.

Overall, ChatGPT is a powerful tool for linguists, with many potential applications in language research, analysis, and processing. As the technology continues to advance, it is likely that we will see even more innovative uses of ChatGPT in the field of linguistics.

Understanding ChatGPT

ChatGPT is a large-scale artificial intelligence language model developed by OpenAI. It is based on the Generative Pre-trained Transformer 3.5 (GPT-3.5) architecture, which is designed to generate natural language text. ChatGPT is capable of performing a variety of language tasks, including language translation, question-answering, text summarization, and more. It has been trained on a massive corpus of text data and uses deep learning algorithms to generate text that is often indistinguishable from human-written text. ChatGPT has the potential to revolutionize many areas of linguistics, including natural language processing, machine translation, and text analysis.

The Role of AI in Linguistics

In linguistics, AI can be used to process large amounts of language data in order to identify patterns, classify languages, and develop models of language learning and use. AI can also be used to assist in language translation, language teaching, and speech recognition.

One of the key advantages of AI in linguistics is its ability to process large amounts of data quickly and accurately. This can help linguists to identify patterns and relationships between different aspects of language use, and to develop more sophisticated models of how language is learned and used.

AI can also be used to automate many of the tasks that linguists traditionally perform manually, such as language transcription and annotation. This can save time and reduce the risk of errors, allowing linguists to focus on more complex tasks.

Another area where AI is having an impact in linguistics is in the development of language translation tools. These tools use machine learning algorithms to analyze large amounts of language data in order to identify patterns and relationships between different languages. This can help to improve the accuracy and speed of language translation, and to make it more accessible to a wider range of people.

Overall, AI is helping to push the boundaries of what we know about language and how it works, and is helping to make language learning and translation more accessible to people around the world.

Chapter 2:
Translation with ChatGPT

ChatGPT can be used for translation tasks, including machine translation, in linguistics. With its ability to understand natural language and generate human-like responses, it can produce translations that are more accurate and natural-sounding than traditional machine translation methods.

ChatGPT can also be used for real-time translation, making it possible to translate live conversations and events in multiple languages. This can be particularly useful for international conferences, meetings, and other events where participants speak different languages.

Furthermore, ChatGPT can help linguists in creating multilingual resources such as dictionaries, terminology lists, and other language resources. It can be trained to recognize patterns and relationships between words and phrases in different languages, making it easier to create accurate and comprehensive resources.

Overall, ChatGPT has the potential to revolutionize the field of linguistics by making translation and language processing faster, more accurate, and more accessible.

Translating Text with AI Assistance

ChatGPT can be an effective tool for translating text between languages. It can be trained on large amounts of parallel texts in different languages and can learn to produce accurate translations. The quality of the translations will depend on the quality and amount of data used to train the model, as well as the complexity of the text being translated.

To translate text with ChatGPT, you can input the text in the source language and use the model to generate the corresponding text in the target language. The output can be further refined by a human translator to ensure accuracy and fluency.

One of the advantages of using AI for translation is the speed at which it can process large volumes of text. This can be particularly useful in situations where a quick turnaround time is required, such as in news reporting or in the translation of business documents. However, it's important to note that AI translation is not always perfect, and it's important to have a human translator review the output to ensure accuracy.

Overall, ChatGPT can be a powerful tool for language professionals looking to streamline their translation workflows and increase efficiency.

Using ChatGPT for Localization

ChatGPT can be used for localization by training it on large amounts of text data in the target language. This will enable the AI model to generate text that is more natural and authentic in the target language.

To use ChatGPT for localization, one can provide it with a large corpus of text in the target language, such as news articles, social media posts, or website content. The model can then be fine-tuned on this data using transfer learning, a process that involves adapting a pre-trained AI model to a new task or domain.

Once the model has been fine-tuned, it can be used to generate text in the target language that is more accurate and natural-sounding than what a non-native speaker could produce. This can be particularly useful for businesses and organizations that need to communicate with a global audience, as it allows them to create localized content that resonates with their target market.

In addition to text translation, ChatGPT can also be used for speech recognition and synthesis, which can further enhance the localization process. By providing audio data in the target language, the model can be trained to recognize and generate speech that sounds more natural and fluent.

Chapter 3:
Interpretation with ChatGPT

Interpretation refers to the process of converting spoken language from one language to another. This can be done consecutively, where the interpreter listens to a speaker and then interprets the speech into the target language, or simultaneously, where the interpreter interprets the speech in real-time as the speaker is talking.

ChatGPT can be used to aid interpretation by providing real-time translation during a conversation. This can be particularly useful in situations where there is a language barrier, such as in international business meetings or conferences. The interpreter can use ChatGPT to quickly translate the speech and then convey it to the audience in the target language.

Additionally, ChatGPT can be used to transcribe speech into text, which can then be translated into another language. This can be particularly useful in situations where there is a need to analyze the content of the speech, such as in legal or medical contexts.

Enhancing Interpretation with AI

AI can enhance interpretation by providing real-time language translation, transcribing speech, and generating summaries. ChatGPT can help in interpreting by providing a tool for quick translations and summaries. It can also be used to transcribe audio or video files to text, making it easier for linguists to work with large amounts of data. Additionally, ChatGPT can assist in identifying nuances in language, such as slang and regional dialects, which can be challenging for human interpreters. This can help ensure that the interpretation is accurate and culturally appropriate. Finally, ChatGPT can help in providing suggestions for the correct terminology or phrasing to use in different contexts, which can improve the quality and consistency of interpretation.

Real-time Interpretation with ChatGPT

Real-time interpretation with ChatGPT refers to using the natural language processing capabilities of ChatGPT to instantly translate spoken language from one language to another during a conversation. This can be particularly useful in situations where people who speak different languages need to communicate with each other, such as in international business meetings or medical consultations.

Real-time interpretation with ChatGPT involves using speech recognition software to transcribe spoken language into text, and then using the language model capabilities of ChatGPT to translate the text into the desired language. The translated text is then spoken out loud using text-to-speech technology.

One of the benefits of using ChatGPT for real-time interpretation is its ability to handle complex sentences and idiomatic expressions that are common in everyday language. However, it is important to note that current AI models are not perfect and may make errors in translation. As with any AI tool, it is important to use ChatGPT as a tool to aid interpretation and not as a replacement for human interpreters.

Chapter 4:
Language Learning with ChatGPT

Language learning is an area where AI and machine learning technologies have significant potential to revolutionize the traditional learning model. ChatGPT can be a useful tool for language learners, as it can generate conversational responses in different languages and provide instant feedback to learners. Here are some ways in which ChatGPT can be used for language learning:

1. Conversational Practice: ChatGPT can generate conversational responses in different languages, allowing learners to practice their speaking and listening skills in a natural and engaging way. This can be especially helpful for learners who do not have access to native speakers or who are unable to travel to countries where their target language is spoken.
2. Writing Practice: ChatGPT can also help learners to practice writing in their target language. It can generate prompts and provide feedback on grammar, vocabulary, and syntax.
3. Vocabulary Building: ChatGPT can be used to generate vocabulary lists and provide explanations and examples of how to use the words in context. It can also suggest synonyms and antonyms to help learners expand their vocabulary.
4. Cultural Learning: ChatGPT can provide learners with cultural context and insight into the use of language in different cultures. It can generate responses that incorporate cultural references and provide explanations of cultural customs and practices.
5. Accent Reduction: ChatGPT can be used to help learners reduce their accent in their target language. It can generate responses and provide feedback on pronunciation, intonation, and stress.
6. Personalized Learning: ChatGPT can be used to create personalized language learning experiences for learners. It can generate content based on learners' interests and learning style, and adapt to their individual progress and needs.

Overall, ChatGPT has the potential to revolutionize the way language learners approach learning a new language. By providing instant feedback and generating engaging content, it can make the language learning process more efficient and enjoyable.

Learning Vocabulary and Grammar with AI

ChatGPT can assist in learning vocabulary and grammar in a foreign language. One way to do this is by providing users with prompts and exercises that help them practice using the language. For example,

ChatGPT can provide sentence prompts and ask users to fill in the blanks with the appropriate word or phrase. It can also ask users to translate sentences from their native language to the foreign language or vice versa.

Another way ChatGPT can assist in language learning is by providing feedback on the user's responses. ChatGPT can highlight errors in grammar or syntax and suggest corrections. It can also provide explanations and examples to help users better understand the rules of the language.

ChatGPT can also provide personalized language learning experiences by adapting to the user's skill level and learning style. It can adjust the difficulty of exercises based on the user's performance and provide additional resources and tips to help users overcome specific challenges they may be facing.

Finally, ChatGPT can also provide immersive language learning experiences by engaging users in conversation in the foreign language. By simulating real-world conversations, ChatGPT can help users develop their listening and speaking skills in a natural and interactive way.

Developing Language Proficiency with ChatGPT

Yes, ChatGPT can assist in developing language proficiency as it can generate writing prompts, provide feedback on writing exercises, and answer language-related questions. Additionally, ChatGPT can simulate conversations in the target language, which can help improve speaking and listening skills. It can also provide real-time translation of spoken language, making it a useful tool for practicing speaking and listening in a foreign language. However, it is important to note that ChatGPT should not be relied on as the sole means of language learning, as it is not a substitute for human interaction and feedback.

Chapter 5:
ChatGPT for Pronunciation and Accent Improvement

ChatGPT can also be used for pronunciation and accent improvement, as it can analyze spoken language and provide feedback on pronunciation accuracy and intonation. This can be especially useful for language learners who want to improve their speaking skills and sound more natural in the target language.

With the help of machine learning algorithms, ChatGPT can identify patterns and common errors in pronunciation and provide personalized feedback on how to improve. It can also generate exercises and drills to help learners practice their pronunciation and intonation.

Furthermore, ChatGPT can also be used to develop conversational skills and improve fluency. By analyzing written or spoken language, it can identify common phrases and expressions, and provide suggestions on how to use them in context. This can be especially helpful for learners who want to sound more natural and confident when speaking the language.

Overall, ChatGPT can be a valuable tool for language learners looking to improve their pronunciation, accent, and conversational skills.

Using AI to Improve Pronunciation and Accent

ChatGPT can be used to improve pronunciation and accent by providing learners with instant feedback and correction. For example, an AI-powered speech recognition system could analyze a learner's pronunciation and accent and suggest improvements in real-time. This can be especially helpful for language learners who do not have access to a native speaker to provide feedback.

AI-powered pronunciation tools can also provide personalized feedback and practice exercises tailored to the learner's needs. These tools can analyze a learner's speech patterns and identify areas where improvement is needed. They can then provide targeted practice exercises to help the learner address those specific areas.

Additionally, ChatGPT can be used to help learners practice their pronunciation and accent in a realistic context. For example, learners can interact with a virtual assistant powered by ChatGPT to practice their conversation skills and receive feedback on their pronunciation and accent. This can be a valuable tool for learners who want to improve their ability to communicate in real-world situations.

Developing Listening Comprehension Skills with ChatGPT

ChatGPT can also be used to develop listening comprehension skills for language learners. With its ability to generate spoken responses, ChatGPT can be used to practice listening and responding in a foreign language. Users can input a spoken sentence or phrase in the target language, and ChatGPT can generate a spoken response in that language for the user to listen to and practice understanding.

Additionally, ChatGPT can also be used to generate audio recordings of texts in the target language. This can be useful for language learners to practice listening comprehension of different accents and dialects.

Chapter 6:
ChatGPT for Language Teaching and Assessment

ChatGPT can be a useful tool for language teaching and assessment. With its ability to generate text and interact with users, it can create a natural language conversation environment that can be used to teach or assess language skills.

Here are some ways ChatGPT can be used in language teaching and assessment:

1. Creating language exercises: Language teachers can use ChatGPT to create exercises for their students. For example, they can provide a prompt and ask the students to respond to it, and then use ChatGPT to generate responses that the students can analyze and evaluate.
2. Providing feedback: ChatGPT can be used to provide feedback to language learners. For example, a learner can provide a sentence or paragraph they have written, and ChatGPT can analyze it and provide feedback on grammar, vocabulary, and other language aspects.
3. Conversational practice: ChatGPT can provide learners with conversational practice. Learners can interact with ChatGPT in a natural language conversation, allowing them to practice their speaking and listening skills.
4. Language assessment: ChatGPT can be used to assess language skills. For example, it can be used to test listening comprehension by playing an audio file and then asking questions about it. It can also be used to test reading comprehension by providing a text and then asking questions about it.

Overall, ChatGPT can be a valuable tool for language teaching and assessment. It can help language learners practice their language skills in a natural and interactive environment, and it can provide teachers with a tool to create language exercises and assess language skills.

Personalized Learning Plans with AI

ChatGPT can assist in developing personalized learning plans that cater to the needs of individual learners. By analyzing data such as a learner's performance, interests, and preferred learning style, ChatGPT can generate tailored lessons and exercises. The AI model can also monitor progress, track achievements, and provide feedback on areas that require improvement.

For example, ChatGPT can generate quizzes and interactive exercises that adapt to the learner's level and pace. It can also create study materials such as flashcards, summaries, and visual aids that cater to the learner's preferred format. Additionally, ChatGPT can monitor the learner's performance and adjust the learning plan accordingly to ensure the best possible outcomes.

Language Assessment and Testing with ChatGPT

Language assessment and testing is an important aspect of language learning and teaching. ChatGPT can assist in this area by providing language proficiency assessment tools based on natural language processing (NLP) and machine learning algorithms.

ChatGPT can analyze language proficiency levels by analyzing written and spoken language samples. This analysis can include identifying grammar and syntax errors, measuring vocabulary size, assessing pronunciation and intonation, and evaluating reading and listening comprehension skills.

ChatGPT can also generate customized language learning plans based on individual learner needs and goals. These plans can include personalized exercises and materials designed to improve areas of weakness and build upon existing strengths.

Furthermore, ChatGPT can facilitate language learning and teaching by providing interactive chatbots and language tutors. These tools can offer personalized feedback and conversation practice, and can simulate real-life language situations to enhance communication skills.

In summary, ChatGPT can support language assessment and testing by providing accurate and personalized proficiency evaluations, and by generating customized learning plans to improve language skills. Additionally, ChatGPT can serve as a language learning and teaching tool by providing interactive chatbots and language tutors to enhance communication skills.

Chapter 7:
Multilingual Chatbots and Customer Service

Multilingual Chatbots and Customer Service are a growing area where ChatGPT can play a significant role. Chatbots can be programmed to engage with customers and provide support in various languages. Chatbots powered by ChatGPT can provide more advanced language capabilities and a more natural language response than traditional chatbots.

For example, a chatbot can be programmed to answer common customer queries in multiple languages, provide product information, or handle transactions in a language of the customer's choice. This can help companies provide better customer service and support, particularly for multilingual customers who may have difficulty communicating in a language they are not proficient in.

Chatbots can also help reduce costs for companies by automating routine tasks and reducing the need for human customer service agents. This can free up agents to handle more complex tasks and provide more personalized service to customers.

However, it's important to note that ChatGPT-powered chatbots are not a replacement for human customer service agents. They can handle routine tasks and answer common queries, but they may not be able to handle more complex issues or provide the same level of empathy and personal touch as a human agent. It's essential to strike a balance between automation and human touch to provide the best possible customer service experience.

Developing Multilingual Chatbots with ChatGPT

ChatGPT can be used to develop multilingual chatbots. Chatbots are computer programs that simulate human conversation and can be used for a variety of purposes, such as customer service, marketing, and information retrieval. With the help of AI, chatbots have become more sophisticated and can now understand natural language input and respond in a more conversational manner.

ChatGPT can be trained on multiple languages and used to develop chatbots that can communicate with users in different languages. This can be particularly useful for businesses that operate in multiple countries or have customers who speak different languages. By providing customer service or other information in the user's preferred language, chatbots can improve the user experience and increase customer satisfaction.

In addition, chatbots can be integrated with machine translation services to provide real-time translation for users who do not speak the language of the chatbot. This can further enhance the chatbot's usefulness and accessibility to a wider audience.

However, it's important to keep in mind that developing a multilingual chatbot requires careful planning and consideration of language nuances, cultural differences, and potential biases. It's essential to ensure that the chatbot is sensitive to different cultures and that it provides accurate and helpful information in all languages it supports.

Improving Customer Service with AI

AI can play a significant role in improving customer service by providing quick and personalized responses to customers' queries and concerns. Chatbots, powered by AI, can handle routine customer queries, such as product information, pricing, and order tracking, freeing up human agents to handle more complex issues. Chatbots can also provide 24/7 customer support, increasing accessibility and responsiveness.

Chatbots can be designed to communicate in multiple languages, making them useful for businesses with a global customer base. Using machine learning, chatbots can also learn from past interactions and improve their responses over time, enhancing the overall customer experience.

Additionally, AI-powered sentiment analysis can be used to track and analyze customer feedback on social media and other platforms, allowing companies to identify and address issues quickly. This can help build customer trust and loyalty, which can lead to increased revenue and growth for the business.

Chapter 8:
ChatGPT for Linguistic Research

ChatGPT can be used in a variety of ways for linguistic research. One of the most significant areas where ChatGPT has been used is in the development of natural language processing (NLP) and natural language understanding (NLU) technologies. These technologies enable computers to process, analyze, and understand human language, which has a wide range of applications in fields such as machine translation, sentiment analysis, and conversational AI.

In addition, ChatGPT can be used in linguistic research to generate text in different languages, which can be used for tasks such as corpus creation, machine translation, and language modeling. ChatGPT can also be used to analyze large amounts of text data and identify patterns or trends, which can be useful in fields such as computational linguistics, sociolinguistics, and discourse analysis.

Moreover, ChatGPT can also be used to create chatbots that simulate human conversation and can be used for language learning or language practice. Finally, ChatGPT can also be used for sentiment analysis, which is the process of analyzing text to determine the emotional tone and sentiment conveyed by the language. This can be useful in fields such as marketing and public opinion research.

Using AI for Linguistic Research and Analysis

AI can be useful in linguistic research and analysis in a number of ways. Here are some examples:

1. Natural Language Processing (NLP): NLP is a field of AI that deals with the interaction between computers and human languages. With NLP techniques, linguists can analyze large datasets of texts, identify patterns, and extract meaningful insights. For example, they can use NLP to analyze the syntax, semantics, and pragmatics of a language, and to detect linguistic features such as tone, sentiment, and register.

2. Corpus Linguistics: Corpus linguistics is a methodology that uses large collections of texts, called corpora, to study language. With AI tools, linguists can create and analyze corpora more easily and quickly than before. For example, they can use machine learning algorithms to automatically classify and tag texts, and to identify patterns of use in a particular language.

3. Speech Recognition and Synthesis: AI can also be used to analyze and synthesize spoken language. For example, linguists can use speech recognition algorithms to transcribe spoken language into written text, and to analyze the phonetics and phonology of a language. They can

also use text-to-speech synthesis to create realistic speech output in a particular language.
4. Language Revitalization: AI can also be used to revitalize endangered languages. For example, linguists can use machine learning algorithms to automatically transcribe and translate spoken language, and to create language learning tools and resources.

Overall, AI can be a valuable tool for linguists in many areas of research and analysis. However, it is important to use AI ethically and responsibly, and to be aware of potential biases and limitations in AI models.

Linguistic Data Mining with ChatGPT

Linguistic data mining involves analyzing large datasets of language use to extract insights and patterns that can be used to better understand language and its use in different contexts. ChatGPT can be used for linguistic data mining tasks such as text classification, sentiment analysis, topic modeling, and named entity recognition.

Text classification is the process of categorizing texts into different predefined classes or categories. For example, one might classify tweets into categories like positive, negative, or neutral sentiments. ChatGPT can be trained on a large dataset of text documents and used to classify new texts based on their content.

Sentiment analysis is a type of text classification that aims to identify the emotional tone of a piece of text. It is commonly used to analyze customer feedback on social media or to track the sentiment of news articles. ChatGPT can be trained to recognize positive, negative, and neutral sentiments in text data.

Topic modeling is a technique used to identify topics or themes in a large collection of texts. It involves analyzing the co-occurrence of words in documents and grouping them into topics based on their similarities. ChatGPT can be used to generate topic models of large datasets of texts, which can then be used to understand the key themes and issues in the corpus.

Named entity recognition is a process of identifying and extracting named entities, such as people, organizations, and locations, from text data. ChatGPT can be trained to recognize named entities in texts, which can then be used to analyze their distribution and frequency across different contexts.

Overall, ChatGPT can be a powerful tool for linguists looking to analyze and understand large datasets of language use. By automating many of the tedious and time-consuming tasks involved in linguistic data mining, ChatGPT can help researchers to identify new insights and patterns in language use more efficiently and effectively.

Chapter 9:
Ethical Considerations in AI-Assisted Linguistics

As with any application of AI, there are ethical considerations that must be taken into account when using ChatGPT for linguistics. Some of the key considerations include:

1. Bias: AI systems can perpetuate and amplify biases present in their training data. When using ChatGPT for linguistics, it is important to be aware of any biases in the training data and take steps to mitigate them.
2. Privacy: ChatGPT relies on large amounts of data, some of which may be sensitive. It is important to ensure that any data used in training and testing the system is obtained and used in an ethical and responsible manner.
3. Transparency: It can be difficult to understand how AI systems like ChatGPT arrive at their conclusions. Transparency and explainability are important considerations in any AI application, and it is important to be able to understand how ChatGPT arrives at its outputs.
4. Responsibility: AI systems like ChatGPT are tools, and it is ultimately the responsibility of the human user to ensure that they are being used in a responsible and ethical manner.
5. Human-in-the-loop: It is important to ensure that human oversight and intervention is included in the use of AI for linguistics. While ChatGPT can be a powerful tool, it is not a replacement for human judgment and expertise.
6. Fairness: It is important to ensure that the use of ChatGPT in linguistics is fair and does not discriminate against any individuals or groups. This can include issues such as language biases, cultural biases, and biases against individuals with certain accents or dialects.

By taking these and other ethical considerations into account, it is possible to use ChatGPT for linguistics in a responsible and beneficial way.

Addressing Bias and Ethics in AI Language Tools

As with any application of artificial intelligence, there are important ethical considerations to take into account when using AI in linguistics. One of the biggest concerns is bias, which can occur in language models due to the inherent biases in the data used to train them. For example, if a language model is trained on text from sources that are primarily written by white men, it may not perform as well when it encounters text written by people from other backgrounds.

To address this, it is important to use diverse and representative datasets when training language models. Additionally, it is important to regularly test and evaluate the performance of these models to identify any biases or errors.

Another important consideration is transparency. It can be difficult to understand exactly how a language model is making its decisions, which can make it challenging to identify and address any biases or errors. To address this, some researchers are working on developing more transparent and explainable AI models.

Finally, there is the question of accountability. As AI language tools become more sophisticated, there is a risk that they may be used to create misleading or harmful content. It is important to have systems in place to ensure that these tools are being used responsibly, and that any negative impacts are identified and addressed.

Ensuring Privacy and Security in Linguistic Data

Protecting privacy and security in linguistic data is critical to the responsible use of AI in linguistics. Linguistic data, which may include personal information, must be kept confidential to protect individuals' privacy.

To ensure privacy and security, it is important to use secure data storage methods and encryption protocols. Access to linguistic data should be strictly controlled, with only authorized personnel being allowed to view or access sensitive information.

Additionally, it is important to obtain informed consent from individuals before using their linguistic data in any research or analysis. This includes providing individuals with information about how their data will be used, who will have access to it, and what steps will be taken to ensure their privacy and security.

Finally, it is important to be transparent about the use of AI in linguistics, including how data is collected, processed, and used. This can help build trust with the public and ensure that individuals are aware of how their linguistic data is being used.

Chapter 10:
Looking Ahead: ChatGPT and the Future of Linguistics

As artificial intelligence and natural language processing technologies continue to evolve, the future of linguistics is likely to be heavily influenced by these advancements. ChatGPT and other language models are already proving to be powerful tools for language translation, interpretation, language learning, and linguistic research.

In the future, it is likely that AI will become even more integrated into our daily lives, enabling more people to communicate with one another across language barriers and breaking down the barriers to communication that have existed for centuries. It is also possible that AI could help researchers gain a better understanding of how languages work, and how they are used in different contexts.

However, as with any new technology, there will be ethical and social considerations that need to be addressed. It is important to ensure that AI language tools are developed and used in ways that are fair, unbiased, and respectful of individual privacy and security. As AI becomes more ubiquitous in our lives, it will be crucial for linguists, AI developers, policymakers, and other stakeholders to work together to address these issues and ensure that the benefits of AI are distributed equitably.

Current Trends and Future Predictions

Recent trends in AI-assisted linguistics include the development of more sophisticated natural language processing (NLP) algorithms, the use of machine learning for language modeling, and the increasing use of AI for speech recognition and synthesis. These advances have led to significant improvements in automated translation, interpretation, and language learning, among other applications.

In the future, it is likely that AI will continue to play an increasingly important role in linguistics, with further improvements in NLP, speech recognition, and language modeling. AI may also be used to develop more advanced language learning tools, including personalized learning plans and virtual tutors.

Another trend is the increasing use of AI in the field of computational linguistics, which involves the study of the computational properties of natural languages. This area of research may lead to the development of new AI applications in fields such as machine translation, text-to-speech synthesis, and speech recognition.

Overall, it is clear that AI will have a significant impact on the field of linguistics in the years to come, both in terms of research and practical applications. However, it is important that these developments are guided by ethical considerations, such as the need to ensure privacy and security in linguistic data, and to address issues of bias and fairness in AI language tools.

Embracing AI in the Linguistic Landscape

As with many other fields, AI is already making an impact in the linguistic landscape and will likely continue to do so in the future. Some current trends and future predictions include:

1. Enhanced language learning: AI-powered language learning tools are becoming more sophisticated, enabling learners to receive personalized feedback and tailored lesson plans. As AI improves, we can expect language learning to become even more efficient and effective.
2. Better translation and interpretation: AI is already improving translation and interpretation capabilities, and this is likely to continue. We may even see more widespread adoption of real-time interpretation tools that use AI to provide more accurate and efficient translation.
3. More multilingual chatbots: As businesses seek to engage with customers in multiple languages, we can expect to see more chatbots that are capable of handling multilingual conversations. This will likely require advances in natural language processing and machine learning.
4. Increased linguistic research: AI is already helping linguists to analyze large amounts of data and identify patterns that might be difficult to detect using traditional research methods. This trend is likely to continue as AI becomes even more sophisticated.
5. Greater ethical considerations: As with any use of AI, it's important to consider the ethical implications of AI-powered linguistic tools. This includes addressing issues of bias, fairness, and privacy, as well as ensuring that linguistic data is used responsibly and ethically.

Overall, AI is likely to play an increasingly important role in the linguistic landscape, enabling us to communicate more effectively across languages and cultures, and helping us to better understand the complexities of language itself.